ROLE OF VITAMIN B_6 IN NEUROBIOLOGY

Advances in Biochemical Psychopharmacology

Volume 4

Advances in Biochemical Psychopharmacology

Series Editors

Erminio Costa, M.D.
Chief, Laboratory of Preclinical Pharmacology
National Institute of Mental Health
Washington, D.C., U.S.A.

Paul Greengard, Ph.D.
Professor of Pharmacology
Yale University School of Medicine
New Haven, Conn., U.S.A.

ROLE OF VITAMIN B_6 IN NEUROBIOLOGY

Advances in Biochemical Psychopharmacology
Volume 4

EDITORS

Manuchair S. Ebadi, Ph. D.
Chairman, Department of Pharmacology
University of Nebraska
College of Medicine
Omaha, Nebraska

Erminio Costa, M. D.
Chief, Laboratory of Preclinical Pharmacology
National Institute of Mental Health
Washington, D.C.

Raven Press · New York · 1972

International Standard Book Number 911216–18–9
Library of Congress catalog card number 73–84113

TO PROFESSOR ESMOND E. SNELL

Preface

As a coenzyme, vitamin B_6 participates in the biotransformation of many amino acids and in the metabolism of proteins, lipids, carbohydrates, purines, pyrimidines, and all compounds implicated as neurotransmitters. It is also involved in the synthesis of five other coenzymes, namely, pantothenic acid, nicotinamide, riboflavin, nicotinamide adenine dinucleotide, and coenzyme A. Vitamin B_6 deficiency is characterized by anemia, growth retardation, and alteration in neuronal function, including neuropathies, hyperirritability, hyperexcitability, and convulsions. The importance of vitamin B_6 in the study of brain function assumes still greater significance when one considers the effects of nutritional deficiencies on the growth and development of the brain and mental processes, as well as the involvement of vitamin B_6 in some inborn errors of metabolism which result in mental retardation.

This volume was organized (a) to review and elaborate on the role of pyridoxal phosphate in the regulation of the metabolism of biogenic amines, (b) to discuss the nature and the mechanism of pyridoxine-responsive seizures, (c) to emphasize the diversity of biochemical reactions involving this coenzyme, and (d) especially to stimulate research on vitamin B_6 metabolism as applied to neurobiological aspects of psychopharmacology. The first three objectives have been fulfilled. It is hoped that this book will herald the apperance of greater interest in research on this coenzyme among students of neurobiology.

Despite the fact that various classes of psychoactive agents alter the level of pyridoxal phosphate, as well as the activities of the B_6-related enzymes and the B_6-dependent enzymes in the brain, no theoretical or experimental explanation of the mechanism of these phenomena has been presented. Although various stimulants of the central nervous system reduce the level of pyridoxal phosphate within the CNS and concomitantly cause hyperexcitability and convulsions, the role that vitamin B_6 plays in the maintenance of the integrity of neuronal excitability is obscure. More research on these and similar effects of vitamin B_6 seems indicated.

Nutritionists have established the various effects and complications of B_6 deficiency, and biochemists have delineated many complex mechanisms of pyridoxal phosphate involvement in B_6-dependent reactions. However, there remains a wealth of knowledge about this coenzyme yet to be discovered by neurophysiologists and neuropathologists who are involved in studies dealing with various neurobiological parameters in health and disease; and by neuropsychopharmacologists who are anxious to study and delineate the pharmacodynamics of various psychoactive agents. It is anticipated that the study of vitamin B_6 in neurobiology will prove most revealing and rewarding.

We are pleased to dedicate this book to Professor Esmond E. Snell, who has defined and described various catalytically active forms of B_6 and has made many outstanding contributions to this field. The generous support of Eaton Laboratories, Hoffmann-La Roche, and Smith, Kline and French is gratefully acknowledged.

December 20, 1971 M. Ebadi and E. Costa

Contents

Advances in Biochemical Psychopharmacology, Vol. 4.
Raven Press, New York © 1972.

Relation of Chemical Structure to Metabolic Activity of Vitamin B_6

Esmond E. Snell

Department of Biochemistry, University of California, Berkeley, California 94720

The relationship of structure to function of physiologically important compounds is a problem central to much of biochemistry and pharmacology, and usually is only poorly understood. This problem arose very early in the history of vitamin B_6, for the observations (a) that pyridoxine was relatively inactive in supporting growth of lactic acid bacteria, but became highly active after appropriate chemical treatments (Snell, 1942, 1944*a*), and (b) that feeding pyridoxine to animals was followed by excretion of compounds much more active than pyridoxine itself in promoting growth of these bacteria (Snell, Guirard, and Williams, 1942) led directly to our discovery of pyridoxal and pyridoxamine (Snell, 1944*a,b*). All three of these compounds are widely distributed in nature and are approximately equally effective in promoting growth of animals on deficient diets. They are now known to function as precursors of pyridoxal 5′-phosphate and pyridoxamine 5′-phosphate which serve as essential coenzymes for a large number of reactions concerned with degradation and biosynthesis of amino acids *in vivo*. Species for which pyridoxine or pyridoxamine show little or no activity fail to convert these compounds to the coenzyme efficiently (for reviews, see Snell, 1958, 1963, and Braunstein, 1960).

We may consider the relationship of structure to activity of these compounds at three levels of increasing specificity: (a) what chemical features of pyridoxal or its phosphate are directly involved in catalysis; (b) what additional structural features are required to permit the coenzyme to interact with its various conjugate apoenzymes to form active holoenzymes; and (c) finally, and most restrictive, what structural features of vitamin B_6 can be varied and still retain vitamin activity in microorganisms or animals? Early experiments in these areas have been reviewed elsewhere (Snell, 1958); here I shall limit the discussion to only a few aspects of this topic, and draw heavily from results published from my own laboratory (with which I am most familiar) for illustrative material.

I. CHEMICAL FEATURES OF PYRIDOXAL REQUIRED FOR CATALYSIS

Delineation of the chemical features of pyridoxal necessary for catalysis was made possible by discovery that this vitamin catalyzed nonenzymatically in dilute aqueous solutions several reactions of amino acids that also occur *in vivo*, where they are catalyzed by enzymes that require pyridoxal phosphate as a coenzyme. Examples of such reactions include transamination between pyridoxal and any of several amino acids to yield pyridoxamine and the corresponding keto acids, as shown in Equations 1 and 2.

$$\text{pyridoxal} + \text{alanine} \rightleftharpoons \text{pyridoxamine} + \text{pyruvate} \quad (1)$$

$$\alpha\text{-ketoglutarate} + \text{pyridoxamine} \rightleftharpoons \text{pyridoxal} + \text{glutamate} \quad (2)$$

$$\textit{Sum:}\ \text{alanine} + \alpha\text{-ketoglutarate} \rightleftharpoons \text{pyruvate} + \text{glutamate} \quad (3)$$

Since these reactions are fully reversible, complete non-enzymatic transamination reactions analogous to that catalyzed, for example, by alanine aminotransferase (Equation 3), will occur on the addition of an appropriate amino acid-keto acid pair; pyridoxal and pyridoxamine appear only as catalysts in such reactions. Similarly, pyridoxal (PL) catalyzes the conversion of many β-substituted amino acids to pyruvate and ammonia, with liberation of the β-substituent, as shown for serine in Equation 4. This reaction is fully analogous to that catalyzed by serine dehydratase *in vivo*.

$$\underset{\displaystyle |\atop \displaystyle NH_2}{HOCH_2CHCOO^-} \xrightarrow[PL]{-H_2O} [\underset{\displaystyle |\atop \displaystyle NH_2}{CH_2 = CCOO^-}] \xrightarrow{+H_2O} CH_3COCOO^- + NH_3 \quad (4)$$

The illustrative data in Table 1 show that serine is completely stable for 10 min at 100° and pH 5, either alone or in the presence of metal (Al^{3+}) ions. On addition of pyridoxal, pyruvate and ammonia are formed slowly in equimolar amounts, and this reaction is greatly speeded (about 20-fold) by the addition of Al^{3+}. For convenience, most non-enzymatic reactions of pyridoxal have been studied at 100°; however, they occur much more slowly at 37°, and, by choosing the appropriate substrates, they can occur quite rapidly at such temperatures. Serine-3-phosphate, for example, in the presence

TABLE 1. *Catalysis of keto acid production from serine by pyridoxal and alum*[a]

Reactants (mM)			Products (mM)	
Serine	Pyridoxal	Alum	Pyridoxal	Pyruvate
10	0	0	—	0.00
10	0	1	—	0.00
10	10	0	9.7	0.23
10	10	1	9.2	5.6
10	pyridoxamine 10	1	0.01	0.04

[a] From Metzler and Snell (1952). The samples were buffered at pH 5.0 with acetic acid (0.2 M)-ammonia buffer prepared from redistilled acid and ammonia and heated for 10 min at 100°.

of pyridoxal is rapidly converted at room temperature to pyruvate, ammonia, and inorganic phosphate, the reaction being speeded about 10-fold by appropriate metal ions (Table 2). Other β-substituted amino acids with a strongly electronegative substituent in the β-position, such as carbamyl serine and azaserine, yield similar results.

By replacing pyridoxal in such reactions by structurally modified compounds, it is possible to define the structural characteristics of the vitamin necessary for effective catalysis of these reactions. The results (Fig. 1) show that the 4-formyl group, the phenolic group, and the heterocyclic nitrogen atom of pyridoxal are essential for the reactions. If the phenolic group is covered by a methyl group, as in 3-O-methylpyridoxal, or if the heterocyclic N is missing, as in salicylaldehyde, no reaction is observed. The 2-methyl group and the 5-hydroxymethyl group are not essential: 3-hydroxypyridine-4-aldehyde and the electronically equivalent *ortho* compound, 3-hydroxy-

TABLE 2. *Comparative rates of pyruvate production from serine-3-phosphate, carbamylserine, and azaserine on incubation with pyridoxal and metal ions*[a]

Time	Pyruvate production (mmoles per liter) at 37° from					
	Serine-3-phosphate		Carbamylserine		Azaserine	
(min)	Complete	No Ga(III)	Complete	No Ga(III)	Complete	No Ga(III)
15	2.0	0.2	3.6	0.3	2.7	0.3
30	2.9	0.2	4.6	0.4	4.6	0.5
60	3.7	0.3	4.9	0.6	5.2	0.8
120	4.8	0.4	6.8	0.7	5.8	1.6

[a] From Longenecker and Snell (1957). The complete reaction mixture contained, per liter, 10 mmoles of serine phosphate, carbamylserine, or azaserine, 2 mmoles of pyridoxal, and 1 mmole of $Ga(NO_3)_3$, and was buffered at pH 9.0 with 0.1 M bicarbonate buffer. No pyruvate was produced in any of the reaction mixtures when pyridoxal was omitted.

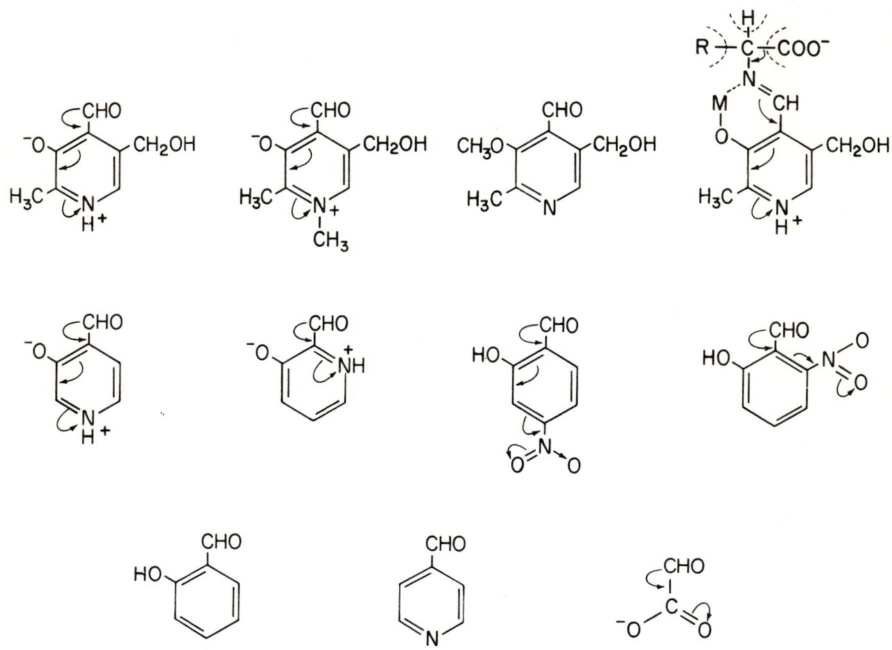

FIG. 1. Structural prerequisites for catalysis of transamination and α, β-elimination reactions of amino acids in aqueous solutions near neutrality. Electron displacements in catalytically active compounds are shown by curved arrows; other compounds are inactive. A metal-ion-stabilized Schiff's base formed between pyridoxal and an amino acid is shown in the upper right. For further discussion, see Snell (1958) and Guirard and Snell (1964).

pyridine-2-aldehyde, are both effective catalysts of such reactions. The role of the pyridine nitrogen as a strongly electronegative center was first suspected from the fact that *p*- and *o*-nitrosalicylaldehyde catalyze somewhat similar reactions of amino acids although much less effectively than pyridoxal.

All of the catalytically active compounds, therefore, contain an acidic hydroxy group *ortho* to the formyl group and a strongly electronegative group so placed as to reduce the electron density about the formyl group. Such compounds react readily with amino acids in aqueous solutions to form Schiff bases which can be stabilized by chelation with a proton or with metal ions, as shown in Fig. 1. In such chelates, the electronegative heterocyclic nitrogen, the azomethine nitrogen, and the chelated metal ion all cooperate to reduce the electron density about the three bonds to the α-carbon atom of the amino acid, thus weakening these bonds and "activating" the amino acid for the several reactions catalyzed by pyridoxal in nonenzymatic reactions.

How such labilization functions to promote dehydration of serine (reaction 4) is clear from Fig. 2. Labilization of the α-hydrogen first occurs followed by elimination of the extra electron pair (present in the first formed

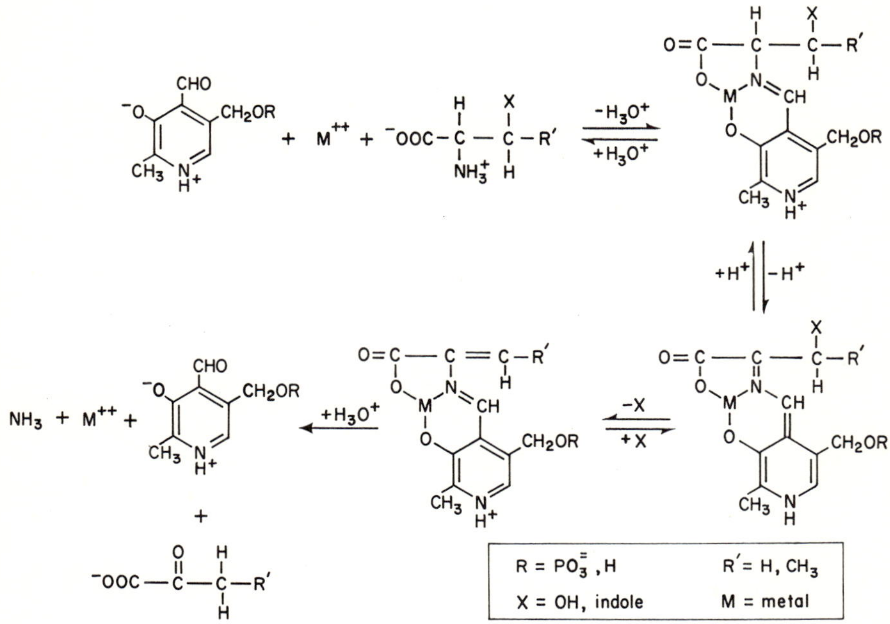

FIG. 2. Mechanism of catalysis of α, β-elimination reactions of serine (R′ = H, X = OH), threonine (R′ = CH_3, X = OH), or tryptophan (R′ = H, X = indolyl residue) by metal ions and pyridoxal (R = H) or pyridoxal-*P* (R = $PO_3^=$).

(quinonoid intermediate) as a hydroxyl ion, with conversion of the quinonoid intermediate to the Schiff's base of α-aminoacrylic acid, which hydrolyzes in aqueous solution to regenerate pyridoxal and metal ions together with free α-aminoacrylic acid. The latter compound is spontaneously unstable in aqueous solution and hydrolyzes to pyruvate and ammonia.

Each of the non-enzymatic reactions catalyzed by pyridoxal can be understood in similar terms, and these reactions appear to be close mechanistic models for the corresponding enzymatic reactions. The subject has been reviewed frequently (e.g., Snell, 1958; Braunstein, 1960; Snell and DiMari, 1970). For our purposes it is important to note that the nonenzymatic reactions can be catalyzed by aldehydes as diverse in structure as pyridoxal and glyoxylate, a common essential feature being the ability to react freely and reversibly with amino acids to form Schiff bases in which appropriately placed electron-attracting groups activate the amino acid by weakening the bonds to its α-carbon atom. We may now compare these chemical requirements for catalysis of non-enzymatic reactions with the structural requirements for coenzyme action.

II. STRUCTURAL FEATURES OF PYRIDOXAL 5′-PHOSPHATE (PLP) NECESSARY FOR PARTICIPATION IN ENZYMATIC REACTIONS

The effect of variations in coenzyme structure on the activity of different PLP-dependent enzymes has been studied recently in several different laboratories; a review is available (Snell, 1970). For our purposes here, a comparison of three enzymes—one that catalyzes a transamination reaction and two that catalyze α, β-elimination reactions—will suffice.

A. Aspartate aminotransferase (AAT).

Catalysis of reaction 5 by this enzyme involves interconversion of pyridoxal-*P* and pyridoxamine-*P* on the enzyme surface as shown in Fig. 3A. Wada and Snell (1962) showed

$$\text{L-aspartate} + \alpha\text{-ketoglutarate} \rightleftharpoons \text{L-glutamate} + \text{oxaloacetate} \qquad (5)$$

that the corresponding apoenzyme could catalyze this same reaction at a very slow rate if free pyridoxal was added, the reaction then proceeding as shown in Fig. 3B. In this latter reaction, pyridoxal obviously occupies the binding site normally occupied by pyridoxal-*P* and is converted to pyridoxamine, just as pyridoxal-*P* in the holoenzyme would be converted to pyridoxamine-*P*. However, the unphosphorylated compounds have comparatively low affinities for the apoenzyme ($<$ 0.001 that of pyridoxal-*P*) and therefore act as diffusible cosubstrates rather than as tightly bound coenzymes. One can safely assume that the 5′-phosphate group, which contributes so

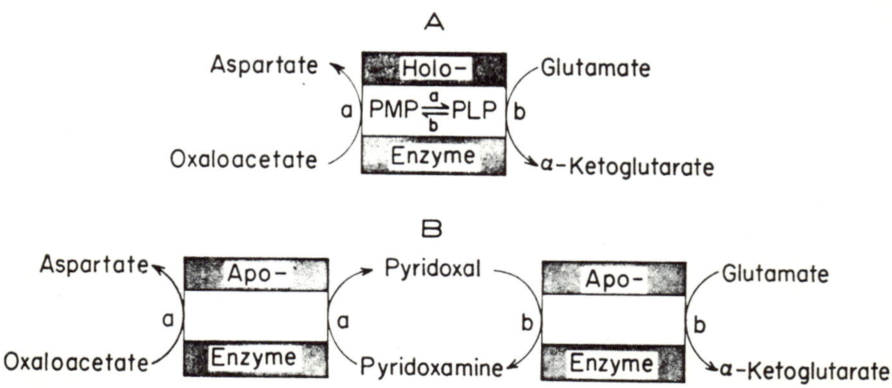

FIG. 3. Schematic representation of reactions catalyzed by (A) aspartate aminotransferase, and (B) the corresponding apoenzyme. Reproduced from Wada and Snell (1962) with consent of the publisher.

greatly to firm binding of the coenzyme, also orients the coenzyme in a position appropriate for further reaction. Both effects promote rapid reaction and are lacking when the unphosphorylated compounds replace the coenzymes. The important point emerges that the 5′-phosphate group of pyridoxal-*P* is required for efficient action of PLP enzymes only because it is required for efficient and conformationally appropriate combination of coenzyme with the apoenzyme; it apparently does not participate in catalysis *per se.*

When pyridoxal-*P* combines with apoAAT, a pronounced spectral band appears at either 430 or 360 nm, depending on the pH (Fig. 4). Similar absorbance changes appear when either 2-nor-PLP, which lacks a methyl group at the 2-position, or 2′-MePLP (ω-MePLP, 2-nor-2-ethyl-PLP), which has an ethyl group at this position, interacts with the apoenzyme. These changes indicate that all three compounds combine in a similar way at the active site of the apoenzyme. The coenzymatic activities of these and related analogues of pyridoxal-*P* are shown in Table 3. Each of the analogues combines with apoAAT as shown both by the spectral changes and the high activity of the resulting complex in catalyzing reaction 5 (*cf.* values for the maximal velocity,

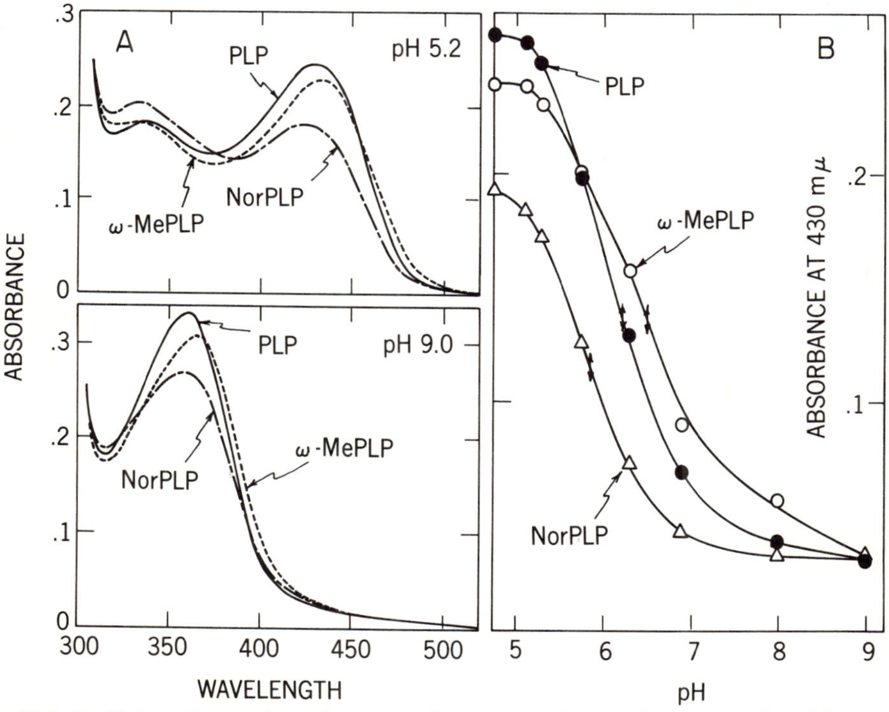

FIG. 4. pH-dependent variation in spectra of aspartate aminotransferase reconstituted from the apoenzyme with PLP, 2-norPLP or 2′-MePLP (ω-MePLP). The arrows (part B) indicate the pK values for the spectral change. For details, see Morino *et al.* (1967). (Reproduced with permission of the publisher.)

TABLE 3. *Some properties of aspartate aminotransferases reconstituted from the apoenzyme with various 6- or 2-substituted analogs of pyridoxal phosphate*[a]

Property measured	Apoenzyme reconstituted with					
	PLP	2-NorPLP	2′-MePLP	2′-Propy1PLP	6-MePLP	2-Nor-6-MePLP
λ_{max}, pH 5.2 (nm)	430	425	435	440	455	455
λ_{max}, pH 8.1 (nm)	360	360	365	370	370	370
pK_a	6.25	5.8	6.5	6.4	6.3	—
V_{max}, relative	1	1.2(1.8)	(0.32)	0.5	0.56	0.47
K_m, α-ketoglutarate (mM)	0.1 (0.16)	0.24(0.5)	(0.05)	0.08	0.14	—
K_m, L-aspartate (mM)	2.0 (3.0)	1.8(5.0)	(0.5)	0.6	1.0	—
K_{co} (μM)[b]	(0.15)	(0.07)	(1.5)	—	—	—

[a] Values in parentheses are from Morino and Snell (1967); all other values are from Bocharov *et al.* (1968).

[b]. Concentration of analog required for half-maximum activity of the reconstituted enzyme under arbitrary nonequilibrium conditions.

V_m, in Table 3). These results demonstrate that neither the 2-methyl group nor the 6-H of pyridoxal-*P* is required for formation of catalytically active complexes with apoAAT: both groups can be replaced either individually (*cf.* 2-nor-PLP, 6-MePLP) or simultaneously (2-nor-6-MePLP) without markedly decreasing activity of the coenzyme. Other studies have shown that pyridoxine-*P* and its 4-deoxy analogue (review, Snell, 1958), as well as 3-O-methylpyridoxal-*P* (Furbish *et al.*, 1969), all of which are inactive in catalyzing non-enzymatic transamination, also do not form catalytically active complexes with apoAAT. N-Methylpyridoxal-*P*, which effectively catalyzes non-enzymatic transamination, is also essentially inactive as a coenzyme for AAT (Furbish, *et al.*, 1969).

We may conclude from these results that activity as a coenzyme requires the same structural features that are required for catalysis of non-enzymatic transamination. However, only some of the compounds that possess these features and catalyze the non-enzymatic reactions fit the coenzyme-binding site of the apoenzyme sufficiently closely to yield catalytically active holoenzyme analogues.

B. Enzymatic α, β-elimination reactions.

The comparative effects of various analogues in replacing pyridoxal-*P* for D-serine dehydratase (which catalyzes reaction 4) and for tryptophanase (which catalyzes the mechanistically similar reaction 6), are summarized in Table 4.

$$\text{L-tryptophan} + H_2O \longrightarrow \text{indole} + \text{pyruvate} + NH_3 \qquad (6)$$

Each active analogue possesses a 4-formyl group, a free 3-phenolic group, and an unsubstituted pyridine nitrogen in the 1-position; i.e., they share the minimum structural features required for catalysis of the corresponding non-enzymatic reactions of Tables 1 and 2. Again, neither the 2- nor the 6-substituent of pyridoxal-*P* is required for activity, although variation in the groups at this position does affect the affinity of the analogue coenzyme for the apoenzyme, as indicated by the K_p and K_{co} values of Table 4, and also affects both the affinity of the resulting analogue holoenzyme for its substrate (as reflected by the K_m values) and the maximal velocity (V_m) of the reaction. It is important to note that these constants vary independently from one enzyme to another. 2′-Hydroxypyridoxal-*P*, for example, combines with aposerine dehydratase almost as well as does pyridoxal-*P*, and the V_m for the analogue holoenzyme is almost twice that of the PLP-apoenzyme complex. By contrast, the affinity of this analogue coenzyme for apotryptophanase is very much less than that of pyridoxal-*P*, and the reconstituted analogue holoenzyme is a less efficient catalyst than is the PLP-enzyme. Even more striking is the result

TABLE 4. *Interaction of pyridoxal phosphate analogues with D-serine apodehydratase and apotryptophanase as determined by activity measurements*

Analogue	D-Serine apodehydratase[a]			Apotryptophanase[b]		
	$K_p^{\bullet}$ (μM)	K_m (mM)	V_{max} (rel)	K_{co} (μM)	K_m (mM)	V_{max} (rel)
A. coenzymatically active						
PLP	0.035	0.37	1.00	0.71	0.27	1.00
2-NorPLP	0.075	0.23	0.80	3.8	0.7	0.6
2′-MePLP	0.11	0.40	0.25	11	0.2	0.04
2′-HOPLP	0.04	0.79	1.85	59	0.22	0.6
6-MePLP	13	0.32	0.20	7.7	0.57	0.2
2-Nor-6-Me PLP	1.3	1.8	0.13	80	2.7	0.06
5′-MePLP	0.035	0.45	0.25		(not tested)	
PL-5′-methylene phosphonate	0.24	1.0	0.30		(not tested)	
B. inhibitory (competitive with PLP)						
2′-Propyl PLP	9.1			20		
2′,2′-DiMePLP	0.23			360		
PL-5′-Me phosphonate	7.2					
PL-5′-(β-cyanoethyl) phosphate	0.95					
PL-5′-acetic acid	0.034					

[a] From Dowhan and Snell (1970). K_p designates apparent dissociation constants for the PLP analogues determined under equilibrium conditions for PLP.

[b] Entries for PLP, 2-NorPLP, and 2′MePLP are from Morino and Snell (1967); other entries are unpublished data of Kagamiyama and Snell. In these cases, K_{co} represents the concentration of PLP analogue necessary to obtain half maximum activity after an arbitrary period of incubation (30 min), or K_i values calculated under similar conditions in the presence of PLP; equilibrium conditions were not defined.

with 2′-propyl-PLP, which combines with apotransaminase to yield an analogue-holoenzyme with high catalytic activity (Table 3). The same compound combines with but does not activate either D-serine apodehydratase or apotryptophanase (Table 4), and competitively inhibits reactivation of these apoenzymes by pyridoxal-*P*. Such variations from enzyme to enzyme show that marked differences exist in the combining site for pyridoxal-*P* in the various enzymes dependent upon this coenzyme. Furthermore, even though an analogue apparently combines normally at the coenzyme binding site (as judged by spectral shifts) and possesses all of the chemical prerequisites for catalytic activity (e.g., 2′-propyl-PLP), binding may be accompanied by spatial distortions within the active site that render the analogue apoenzyme complex inactive as a catalyst. Such minor distortions within the active site appear to be the most likely explanation for the catalytic inactivity of analogue holoenzymes reconstituted with analogues such as pyridoxal 5′-acetic acid and pyridoxal 5′-methylphosphonate, which differ slightly from pyridoxal-*P* both in the size and, more importantly, in the charge of the side chain at position 5.

With each of these three enzymes, an active coenzyme must possess the minimal structural features necessary for catalysis of the corresponding nonenzymatic reactions; each enzyme, however, imposes its own additional structural requirements. These additional strictures stem from the requirement that the coenzyme combine with the binding site of the apoenzyme and do so in such a way that an active site of specific conformation, with narrowly defined spatial tolerances, is produced.

III. STRUCTURAL FEATURES OF VITAMIN B_6 REQUIRED FOR VITAMIN ACTIVITY

To show vitamin B_6 activity, a compound must be absorbed, then transformed into pyridoxal-*P* or into an analogue coenzyme that replaces pyridoxal-*P* for each of the many enzymatic processes dependent upon this coenzyme. Both absorption and transformation into coenzyme are enzymatic processes which impose their own special structural requirements upon their substrates (review, Snell, 1958); furthermore, since the coenzymatic specificity of individual apoenzymes differs (see preceding section), the necessity for meeting each of these many different specificity requirements with a single compound sets much more stringent limits upon the structure of that compound than those set by any single apoenzyme. We may now inquire if *any* variation in the structure of the known forms of vitamin B_6 can be made and still retain vitamin activity.

Excluding trivial variations (such as esterification, etc.) which permit reconversion *in vivo* to vitamin B_6, the enzymatic results discussed in the preceding section suggest that the only structural variants with any chance of possessing vitamin activity would comprise compounds varying in the substituents at the 2- and 6-positions, respectively: minor alterations at these positions do not render coenzyme analogues inactive for any of the enzymes so far tested. Of these possible candidates, only 2′-methylpyridoxine has been tested in any adequate fashion; a few results also are available for 2-norpyridoxine.

A. Vitamin activity of 2′-methylpyridoxal for *Streptococcus faecalis*.

2′-Methylpyridoxine inhibits growth of *Ceratosomella ulmi* (Robbins and Ma, 1942) and of *Saccharomyces carlsbergensis* (Rabinowitz and Snell, 1953). In *Streptococcus faecalis*, a more interesting result was obtained. This organism and similar lactic acid bacteria can be grown in a medium which contains glucose, appropriate vitamins and mineral salts, and all of the L-amino acids provided that either D-alanine or pyridoxal is added (Table 5). Under these conditions, 2′-methylpyridoxal effectively replaces pyridoxal as a growth factor. When the organism was grown with either D-alanine or pyridoxal, the cells contained D-alanine; however, when the cells were grown with D-alanine, the organism contained no vitamin B_6 (Holden and Snell, 1949). We interpreted this result to mean that the sole function for which external supplies of vitamin B_6 was required under these conditions was synthesis of D-alanine [which was subsequently shown to be required for cell wall formation (Snell, Radin, and Ikawa, 1955)]. When D-alanine was supplied preformed, the necessity for this reaction was bypassed and the organism grew without external supplies of vitamin B_6. Since 2′-methylpyridoxal replaces pyridoxal for growth under these conditions, it must be converted intracellularly to 2′-methylpyridoxal-*P*, and this analogue coenzyme must serve as an effective coenzyme for the enzyme that synthesizes D-alanine. D-alanine is known to be synthesized under these conditions from L-alanine through action of the pyridoxal phosphate-dependent alanine racemase, and Fig. 5 shows that 2′-methylpyridoxal-*P* does in fact replace pyridoxal-*P* for this enzyme.

If D-alanine is supplied under the conditions of Table 5 and any of the

TABLE 5. *Comparative growth response of S. faecalis to pyridoxal, 2′-methylpyridoxal, and D-alanine*[a]

	Supplement						
	Pyridoxal			2′-Methylpyridoxal		D-Alanine	
nmoles per 6 ml	0	0.003	0.010	0.10	0.25	4000	6000
dry cells per ml, *mg*[b]	0	0.72	1.2	0.76	1.1	0.64	1.2

[a] From Olivard and Snell (1955*a*). [b] After 15-hr incubation at 37°.

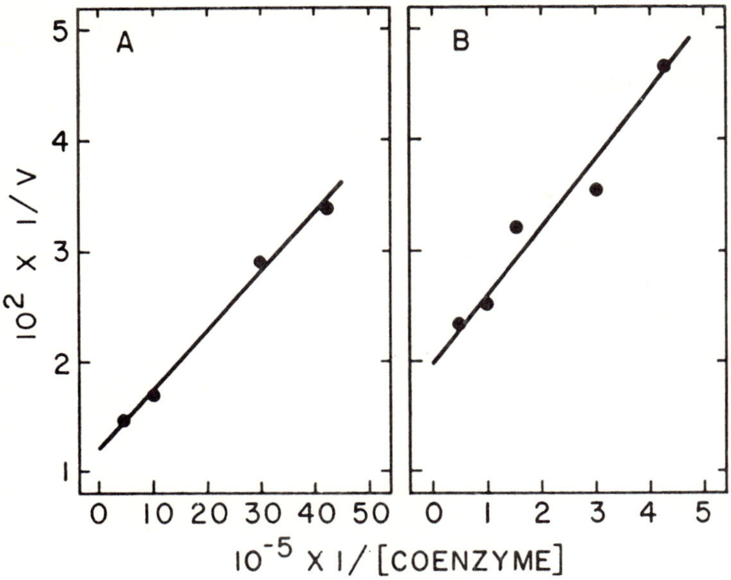

FIG 5. The relation of concentration of pyridoxal-*P* (A) or 2′-methylpyridoxal-*P* (B) to velocity of racemization of L-alanine by cell-free preparations of alanine aporacemase from *S. faecalis*. See Olivard and Snell (1955*a*) for details. (Reproduced with permission of the publisher.)

L-amino acids are omitted, vitamin B_6 is again required to permit synthesis of the missing L-amino acid. This circumstance permits a comparison of the growth-promoting activity of vitamin B_6 analogues under conditions where a single vitamin B_6-dependent reaction limits growth with the ability of that same analogue (or its 5′-phosphate) to activate the enzyme catalyst required for that reaction. For example, if D-alanine is present with all of the L-amino acids except L-leucine (the carbon skeleton of leucine being supplied as α-ketoisocaproic acid), then either pyridoxamine or 2′-methylpyridoxamine permits growth (Fig. 6). Since leucine is formed by transamination of the corresponding keto acid, we should predict from this result that the leucine-α-ketoglutarate aminotransferase should be activated by 2′-methyl-PLP, and this result is indeed obtained (Fig. 7), even though the analogue coenzyme is substantially less active in this capacity than is pyridoxal-*P*. In a similar fashion, 2′-methylpyridoxamine permits growth in the absence of L-phenylalanine and L-tyrosine, and 2′-methylpyridoxal-*P* activates the corresponding cell-free transaminases (Table 6).

In each of these cases, 2′-methylpyridoxamine serves as a complete substitute for pyridoxamine for growth through unlimited transfers. Nevertheless, it is not a substitute for the vitamin under all conditions. For example, *S. faecalis* grows in the absence of L-cysteine when small amounts of py-

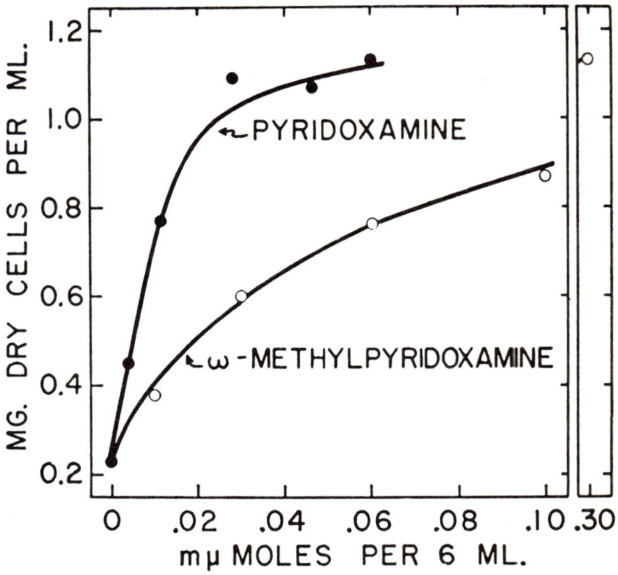

FIG. 6. Comparative activities of pyridoxamine-*P* and 2′-methylpyridoxamine-*P* in promoting growth of *S. faecalis* in a leucine-free medium containing α-ketoisocaproate. Reproduced from Olivard and Snell (1955*b*) with permission of the publisher.

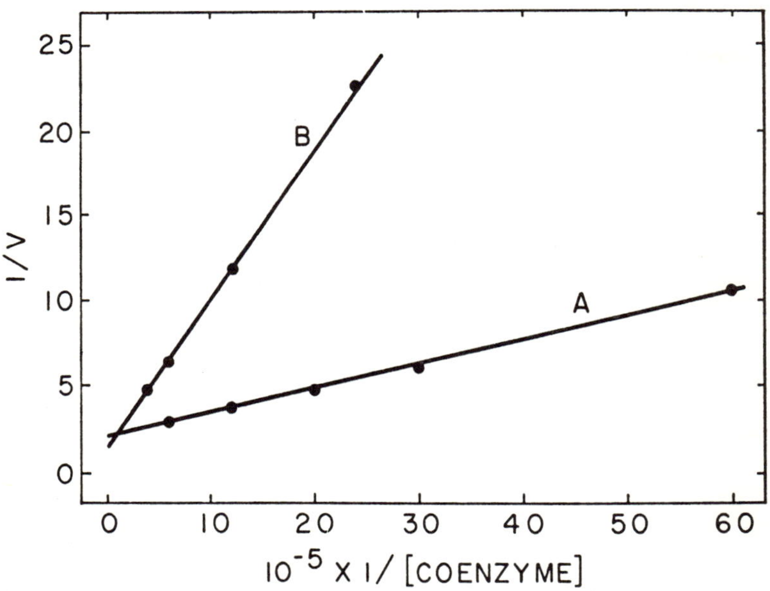

FIG. 7. Comparative activation of leucine apoaminotransferase from *S. faecalis* by pyridoxal-*P* (Curve A) and by 2′-methylpyridoxal-*P* (Curve B). From Olivard and Snell (1955*b*).

TABLE 6. *Comparative growth and enzyme-activating activities of Vitamin B_6 and its 2′-methyl analogues in S. faecalis*[a]

Growth promotion				Enzyme activation	
Growth limited by synthesis of	PMP required for half-maximal growth (nmole per 6 ml)	Ratio of amounts required for activity[b] PMP:2′-MePMP	Growth-limiting enzyme	K_{co} for PLP (μM X 10^6)	Ratio of amounts required for activity[b] PLP:2′-MePLP
D-alanine	0.0024	1:11–16	alanine racemase	0.44	1:10
L-leucine	0.009	1:4.6	leucine-glutamic acid transaminase	1.2	1:10
L-phenylalanine	0.008	1:11	phenylalanine-glutamic acid transaminase	9.0	1:1
L-tyrosine	0.020	1:7.3	tyrosine-glutamic acid transaminase	7.1	1:1.3

[a] From Olivard and Snell (1957*b*).
[b] Comparison made from amounts required for half-maximal growth.

ridoxamine are supplied (Fig. 8), but not when 2′-methylpyridoxamine is supplied instead; indeed, 2′-methylpyridoxamine competitively inhibited growth under these conditions. Thus 2′-methylpyridoxamine cannot replace pyridoxamine for growth in the absence of L-cysteine and would therefore presumably fail to activate the unidentified enzyme that is required in this instance. 2′-Methylpyridoxal-*P* also failed to activate the cysteine dehydratase of this organism (Olivard and Snell, 1955*b*).

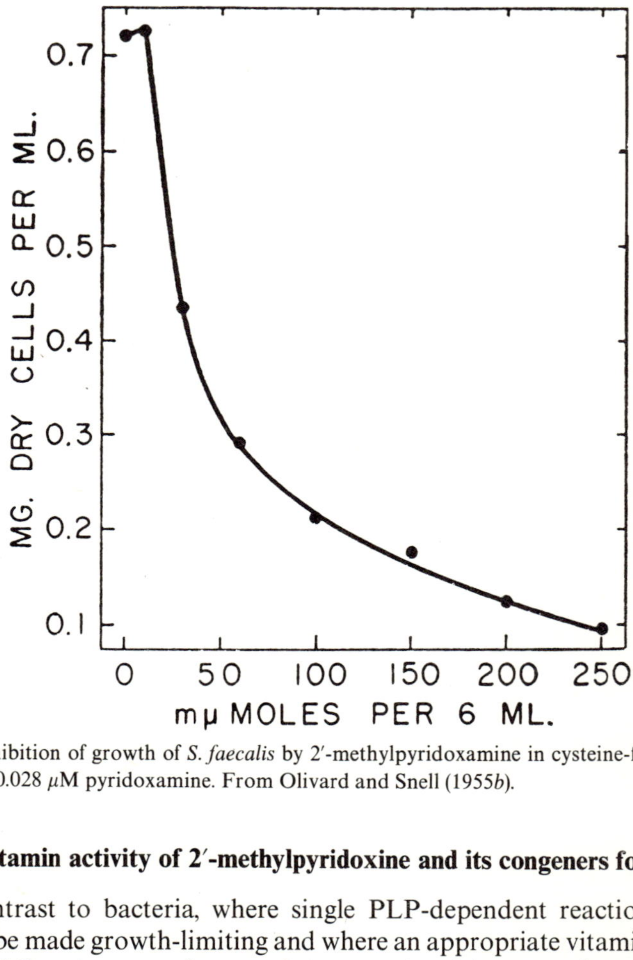

FIG. 8. Inhibition of growth of *S. faecalis* by 2′-methylpyridoxamine in cysteine-free medium containing 0.028 μM pyridoxamine. From Olivard and Snell (1955*b*).

B. Vitamin activity of 2′-methylpyridoxine and its congeners for rats.

In contrast to bacteria, where single PLP-dependent reactions can occasionally be made growth-limiting and where an appropriate vitamin analogue provides a full replacement for permitting certain of these reactions to proceed, growth of animals on vitamin-restricted diets is more complex. In rats, for example, pyridoxal-*P* (or its precursor vitamins) is required even on diets containing a complete assortment of dietary amino acids, to permit their degradation or conversion to other essential intermediary metabolites, to permit normal glycogen metabolism, etc. It is interesting, therefore, that 2′-methylpyridoxal and its congeners appear to replace pyridoxal as a growth factor for rats on a vitamin B_6-deficient ration when the activities of the two compounds are compared over a standard 4-week interval (Table 7). At low levels, 2′-methylpyridoxal continues to permit growth for at least 12 weeks (Fig. 9). However, higher levels of the same compound permitted growth for only 4 weeks, after which growth on the supplemented ration was no better than that on the vitamin-free diet. Sound-induced convulsions are not a prominent symptom in the animals fed this vitamin B_6-deficient ration, and only about 10% of the animals showed such convulsions after 10–12 weeks on the diet. However, in the groups of animals fed 2′-methylpyridoxal, such convulsions occurred in 100% of the animals after only about 4 weeks on the supplemented diet. We conclude that the analogue is not a complete substitute for vitamin B_6 on this ration, and that the deficiency induced by

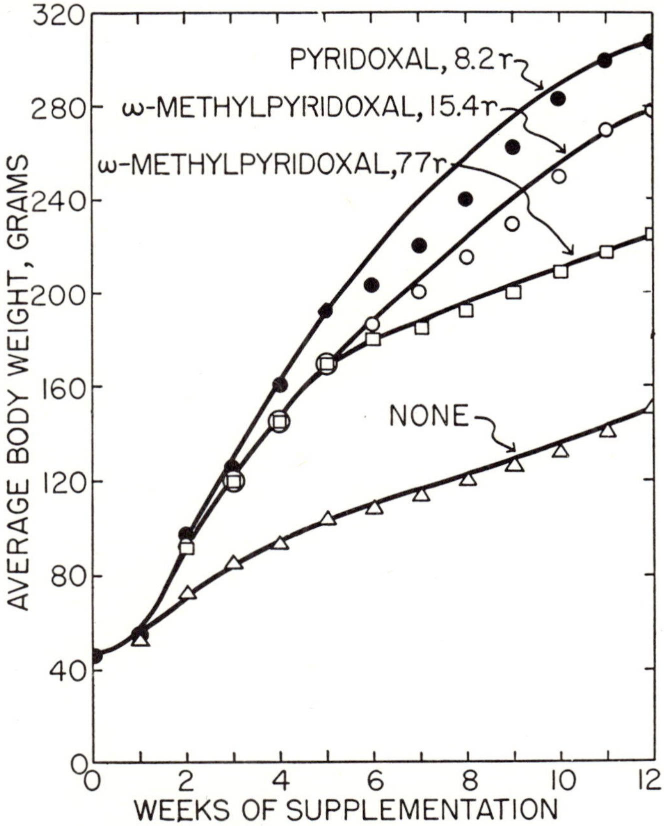

FIG. 9. Comparative growth-promoting effects of pyridoxal and 2′-methylpyridoxal (ω-methylpyridoxal) during long-term administration to rats. The indicated supplements were administered daily. Reproduced from Sandman and Snell (1955) with consent of the publisher.

TABLE 7. *Comparative effects of pyridoxal and 2′-methyl analogues of vitamin B_6 on growth of rats. Each group contained six animals.*[a]

Group	Supplement	Amt. of supplement/ day (μg)	Wt gain in 4 wk (g)	Apparent molar activity (pyridoxal = 100)
1	0	—	46	
2	pyridoxal	1.6	71	
3	pyridoxal	4.1	94	
4	pyridoxal	8.2	114	
5	2′-methylpyridoxal	16.6	100	32
6	2′-methylpyridoxal	83	100	6.5
7	2′-methylpyridoxamine	17.8	92	23
8	2′-methylpyridoxamine	88	101	6.4
9	2′-methylpyridoxine	16.6	100	32
10	2′-methylpyridoxine	83	93	5.1
11	pyridoxal +	82	140	
	2′-methylpyridoxal	83		
12	pyridoxal +	82	134	
	2′-methylpyridoxamine	88		
13	pyridoxal +	82	126	
	2′-methylpyridoxine	83		

[a] From Sandman and Snell (1955).

TABLE 8. *Effects of 2′-methylpyridoxine on growth and survival of vitamin B_6-deficient rats. Each group contained five animals.*[a]

Supplement	Amt. fed/ day (μg)	Avg. wt. gain (g)/wk during					Survivors at 41 days
		1st	2nd	3rd	4th	5th wk	
0		3.2	1.3	0.9	3.3	0.5	5
pyridoxine	8.2	27	16	15	16	8.3	5
2′-methylpyridoxine	820	28	7.9	1.0	0.1	−10.5	1

[a] From Sandman and Snell (1955).

its presence differs from that produced by simple absence of the vitamin from the ration. Very high levels of the analogue (Table 8) produced normal growth responses during the first week of feeding, but became progressively less effective in comparison to pyridoxal; after the third week, the analogue suppressed growth below that observed on the vitamin B_6-deficient ration. In the fifth week, the analogue-fed animals lost weight and died whereas those on the unsupplemented deficient ration, although they failed to grow, remained alive. All of these toxic effects of high levels of 2′-methylpyridoxine could be prevented by simultaneously feeding pyridoxine.

Although the explanation for these unusual results is not clear, the results with the same analogue in isolated enzyme systems and in growing bacteria suggest that at least two different effects are involved. (1) The unphosphorylated analogue is converted to 2′-methylpyridoxal-*P*, which may replace pyridoxal-*P* for some enzymes but not for all, just as it did in the bacterial systems. The relative activities of different vitamin B_6-dependent enzymes would then differ from those obtained on the deficient diet alone, perhaps explaining the prominence of symptoms such as convulsions in the analogue-fed animals. (2) In addition, the analogue coenzyme may displace pyridoxal-*P* from enzymes such as phosphorylase and thus make it available for reactions which the analogue coenzyme cannot itself catalyze. Over a limited time, the analogue-fed animal would thus grow with less than the normally required amount of vitamin B_6. Over a longer period, however, displacement of pyridoxal-*P* from enzymes not activated by 2′-methylpyridoxal-*P* would occur; the animals would then lose weight and die more rapidly than the control animals to which no displacing agent was given.

C. Effects of 2-norpyridoxal and its congeners in microorganisms and animals.

2-Norpyridoxal-*P* is a more effective substitute for pyridoxal-*P* than 2′-methylpyridoxal-*P* in every enzyme so far tested (see Snell, 1970, and section II). It is surprising, therefore, that 2-norpyridoxine is inert as a substitute for pyridoxine in rats (Pol and Klein Obink, 1961) and yeast (Pol and Klein Obink, 1961; Gadekar *et al.*, 1962). 2-Norpyridoxal is also essentially inactive in supporting growth of *S. faecalis* and is only slightly more effective for *Lactobacillus casei* (Table 9); under these same conditions, 2′-methylpyridoxal is substantially more effective in supporting growth (*cf.* Fig. 6). The order of effectiveness of the two unphosphorylated analogues in supporting growth is thus the reverse of that of their phosphorylated forms in enzyme activation (see, e.g., Table 5). This must reflect inability of these test organisms to absorb and/or transform 2-norpyridoxal to 2-norpyridoxal-*P*; unfortunately, the comparative activities of these analogues as substrates for pyridoxal kinase have not been determined.

TABLE 9. *Comparative growth-promoting properties of pyridoxal and 2-norpyridoxal for lactic acid bacteria*[a]

Amino acid omitted from complete medium[b]	*S. faecalis*		*L. casei*	
	PL	2-nor-PL	PL	2-nor-PL
	(ng per 6 ml required for half-maximum growth)[c]			
L-alanine	7.5	– (300)	6.0	120
L-leucine	7.5	– (300)	1.5	130
L-methionine	7.5		11.7	300
L-phenylalanine	6.4	> 300	3.3	90
L-serine	3.1	– (300)		
L-valine	12	– (300)	3.0	300

[a] Unpublished data from B. M. Guirard and E. E. Snell. The basal medium and general growth procedures were similar to those described by Kihara *et al.* (1952) for *S. faecalis;* for *L. casei*, this same amino acid mixture was used with the medium of Kihara and Snell (1960).

[b] The carbon skeletons of the amino acids cannot be synthesized by these organisms in some cases, and were supplied as follows: for L-Ala, D-Ala; for L-Leu, α-ketoisocaproate; for L-Met, no substitution; for L-Phe, phenylpyruvate; for L-Ser, no substitution; for L-Val, α-ketoisovalerate.

[c] A minus sign indicates that no growth was obtained at the highest concentration (in parentheses) of the analogue tested.

IV. CONCLUDING REMARKS

In summary, we have shown that for catalytic activity in non-enzymatic model reactions, the groups at positions 2, 5, and 6 of pyridoxal-*P* (structure A) are not required; the free phenolic group, the formyl group, and the hetero-

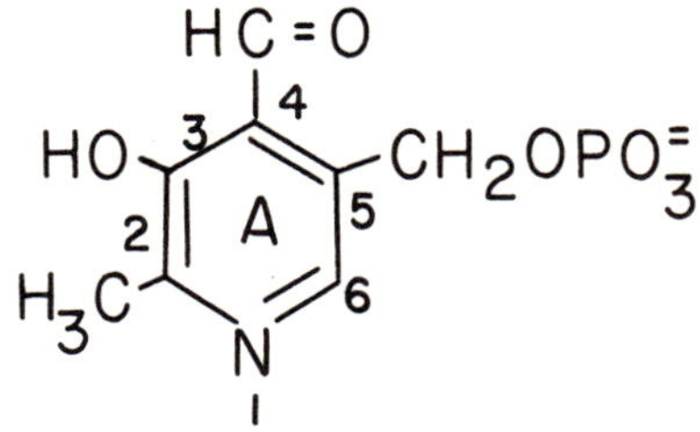

cyclic nitrogen atom at positions 3, 4, and 1 are required, although the nitrogen may be methylated and still retain activity. For high coenzymatic activity, structural requirements are more demanding. The groups at positions 2 and 6 are again not required for activity; unmodified groupings at positions 1, 3, and 4 are now required for high activity. The phosphate residue at position 5′ also is required for high activity, although apparently only to increase the affinity of the coenzyme for the apoenzyme and to assure its proper positioning on the enzyme surface. The permissible structural variations in the coenzyme vary to some extent from enzyme to enzyme, presumably reflecting differences in the combining sites at the enzyme surface. Because of this variation and the specific structural requirements for conversion to pyridoxal-*P*, no modification in structure of vitamin B_6 can be made with retention of full vitamin activity for any organism so far tested. Under restricted conditions of testing, however, 2′-methylpyridoxal can replace vitamin B_6 for certain lactic acid bacteria, and over short periods of time only, for rats.

REFERENCES

Bocharov, A. L., Ivanov, V. I., Karpeisky, M. Ya., Mamaeva, O. K., and Florentiev, V. L. (1968): The characteristics of alkyl substituted analogues of pyridoxal phosphate as coenzymes for aspartate transaminase. *Biochemical and Biophysical Research Communications, 30*:459.

Braunstein, A. E. (1960): Pyridoxal phosphate. In: *The Enzymes*, 2nd Ed., Vol. 2, edited by P. D. Boyer, H. Lardy, and K. Myrback, pp. 113–184. Academic Press, New York.

Dowhan, W., Jr., and Snell, E. E. (1970): D-Serine dehydratase from *Escherichia coli.* III. Resolution of pyridoxal 5′-phosphate and coenzyme specificity. *Journal of Biological Chemistry, 245*:4629.

Furbish, F. S., Fonda, M., and Metzler, D. E. (1969): Reaction of apoaspartate aminotransferase with analogs of pyridoxal phosphate. *Biochemistry, 8*:5169.

Gadeker, S. M., Frederick, J. L., and DeRenzo, E. C. (1962): Synthesis and biological activity of pyridoxine analogs. *Journal of Medicinal and Pharmaceutical Chemistry, 5*:531.

Guirard, B. M., and Snell, E. E. (1964): Vitamin B_6 function in transamination and decarboxylation reactions. In: *Comprehensive Biochemistry*, Vol. 15, edited by M. Florkin and E. H. Stotz, pp. 138–199. Elsevier Publ. Co., Amsterdam.

Holden, J. T., and Snell, E. E. (1949): The vitamin B_6 group. XVII. The relation of D-alanine and vitamin B_6 to growth of lactic acid bacteria. *Journal of Biological Chemistry, 178*:799.

Kihara, H., Klatt, O. A., and Snell, E. E. (1952): Peptides and bacterial growth. III. Utilization of tyrosine and tyrosine peptides by *Streptococcus faecalis. Journal of Biological Chemistry, 194*:801.

Kihara, H., and Snell, E. E. (1960): Peptides and bacterial growth. VIII. The nature of strepogenin. *Journal of Biological Chemistry, 235*:1409.

Longenecker, J. B., and Snell, E. E. (1957): Pyridoxal and metal ion catalysis of α, β-elimination reactions of serine-3-phosphate and related compounds. *Journal of Biological Chemistry, 225*:409.

Metzler, D. E., and Snell, E. E. (1952): Deamination of serine. I. Catalytic deamination of serine and cysteine by pyridoxal and metal salts. *Journal of Biological Chemistry, 198*:353.

Morino, Y., and Snell, E. E. (1967): Coenzymatic activity of homologues of pyridoxal phosphate. *Proceedings of the National Academy of Sciences (U.S.), 57*:1692.

Olivard, J., and Snell, E. E. (1955*a*): Growth and enzymatic activities of vitamin B_6 analogues. I. D-Alanine synthesis. *Journal of Biological Chemistry, 213*:203.

Olivard, J., and Snell, E. E. (1955*b*): Growth and enzymatic activities of vitamin B_6 analogues. II. Synthesis of miscellaneous amino acids. *Journal of Biological Chemistry, 213*:215.

Pol, G., and Klein Obbink, H. J. (1961): Tests of the biological activity of nor-vitamin B_6. *Recueil des Travaux Chimiques, 80*:217.

Rabinowitz, J. C., and Snell, E. E. (1953): Vitamin B_6 antagonists and growth of microorganisms. II. 5-Desoxypyridoxal and related compounds. *Archives of Biochemistry and Biophysics, 43*:408.

Robbins, W. J., and Ma, R. (1942): Specificity of pyridoxine for *Ceratostomella ulmi. Bulletin of the Torrey Botanical Club, 69*:342.

Sandman, R., and Snell, E. E. (1955): Some effects of ω-methyl analogues of vitamin B_6 in rats. *Proceedings of the Society for Experimental Biology and Medicine, 90*:63.

Snell, E. E. (1942): Effect of heat sterilization on growth-promoting activity of pyridoxine for *Streptococcus lactis* R. *Proceedings of the Society for Experimental Biology and Medicine, 51*:356.

Snell, E. E. (1944*a*): The vitamin B_6 group. I. Formation of additional members from pyridoxine and evidence concerning their structure. *Journal of the American Chemical Society, 66*:2082.

Snell, E. E. (1944*b*): The vitamin activities of "pyridoxal" and "pyridoxamine." *Journal of Biological Chemistry, 154*:313.

Snell, E. E. (1958): Chemical structure in relation to biological activities of vitamin B_6. *Vitamins and Hormones, 16*:77.

Snell, E. E. (1963): Some enzymatic transformations of vitamin B_6 and their inhibition. In: *Molecular Basis of Enzyme Action and Inhibition (Proceedings of the Fifth International Congress of Biochemistry, Moscow),* Vol. 4, edited by P.A.E. Desnuelle, pp. 268–279. Pergamon Press, New York.

Snell, E. E. (1970): Analogs of pyridoxal or pyridoxal phosphate: Relation of structure to binding with apoenzymes and to catalytic activity. *Vitamins and Hormones, 28*:265.

Snell, E. E., and DiMari, S. (1970): Schiff base intermediates in enzyme catalysis. In: *The Enzymes,* 3rd Ed., Vol. 2., edited by P.D. Boyer, pp. 335–370. Academic Press, New York.

Snell, E. E., Guirard, B. M., and Williams, R. J. (1942): Occurrence in natural products of a physiologically active metabolite of pyridoxine. *Journal of Biological Chemistry, 143*:519.

Snell, E. E., Radin, N., and Ikawa, M. (1955): The nature of D-alanine in lactic acid bacteria. *Journal of Biological Chemistry, 217*:803.

Wada, H., and Snell, E. E. (1962): Enzymatic transamination of pyridoxamine. *Journal of Biological Chemistry, 237*:127.

Advances in Biochemical Psychopharmacology, Vol. 4.

The Role of Pyridoxal 5′-Phosphate in the Structure and Function of Glycogen Phosphorylase

Jerry L. Hedrick

Department of Biochemistry and Biophysics, University of California, Davis, California 95616

The presence of pyridoxal 5′-phosphate (PLP) in glycogen phosphorylase (E.C.2.4.1.1., α-1,4-glucan: orthophosphate glucosyl transferase) was discovered some 13 years ago; however, in spite of efforts by many investigators since that time, its role is still incompletely understood. This article will attempt to summarize the various hypotheses as to the roles that have been postulated for PLP in phosphorylase, to assess the major evidence in support of these hypotheses, and to comment on the direction future experimentation might take to further elucidate the role of PLP in this enzyme. This article will not be an attempt at a comprehensive review of the published literature on phosphorylase, but rather an evaluation of where we are and where we are going with regard to PLP in phosphorylase. The reader should consult other reviews on phosphorylase which cover various aspects of this multifaceted enzyme (Fischer and Krebs, 1966; Krebs and Fischer, 1964; Fischer, 1964; Fischer *et al.*, 1963; Brown and Cori, 1961).

I. THE FINDING OF PLP IN PHOSPHORYLASE

The first suggestion that phosphorylase contained PLP was made by Velick and Wicks in 1951, who employed microbiological assays for the vitamin. However, PLP was not considered a prosthetic group at that time because it could apparently be removed by repeated crystallization of the enzyme without affecting enzymatic activity.

In the late 1950's, two laboratories independently confirmed the existence of covalently bound PLP in phosphorylase by chemical means, showed that it was required for the functioning of the enzyme, and set the direction of research for subsequent investigations (Baranowski *et al.*, 1957; Cori and Illingworth, 1957; Kent *et al.*, 1958). PLP has since been found to be present in every glycogen phosphorylase described, regardless of its tissue source or the organism from which it has been isolated (Krebs and Fischer, 1964). Thus its universal distribution attests to its essentiality for the functioning of phosphorylase. From the initial observations referred to above came two hypotheses about the function of PLP in phosphorylase: (1) that it played a catalytic role in the physical-organic chemistry sense in the phosphorolysis of glycogen, and (2) that it played a structural role in maintaining the protein in an enzymatically active conformation. More recently, two additional functions have been suggested: (3) that the vitamin might function in controlling the activity of the enzyme (Shaltiel *et al.*, 1966), and (4) that phosphorylase may serve as a depot or reserve storage of PLP (Krebs and Fischer, 1964). Before discussing these hypotheses in detail, let us first examine some of the more salient and relevant features of this complex enzyme.

II. THE INTERCONVERSION OF INACTIVE AND ACTIVE FORMS AND THE STRUCTURE OF RABBIT MUSCLE PHOSPHORYLASE

Although glycogen phosphorylases[1] from many different tissues and organisms have been described, the vast majority of the published work has been carried out on the rabbit muscle enzyme. It exists in two interconvertible forms—phosphorylase *b*, the form inactive in the absence of AMP, and phosphorylase *a*, the form active in the absence of AMP. The interconversion of

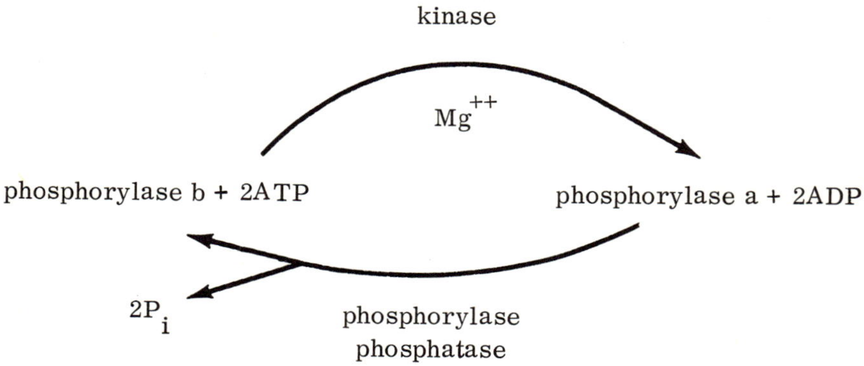

[1] The term phosphorylase will be used as a general term referring to characteristics shared by phosphorylase *a* and phosphorylase *b*.

the two enzyme forms is controlled by an intricate system of enzymes in a "cascading" type of arrangement that also involves active and inactive forms. The "trigger process" which activates this control system that results in the ultimate conversion of phosphorylase *b* to phosphorylase *a* and the initiation of glycogenolysis is either hormonal (epinephrine) or neural in nature (Fig. 1). Chemically, the interconversion of the two phosphorylase forms involves the phosphorylation or dephosphorylation of the protein, which is mediated by a phosphorylase kinase or a phosphorylase phosphatase.

Phosphorylase *a* and *b* are polymeric proteins as originally shown by Madsen and Cori (1956). More recently, the molecular weight of the protein subunit has been reevaluated and is approximately 90,000 g/mole (DeVincenzi and Hedrick, 1967; Seery *et al.*, 1967; Ullman *et al.*, 1968; Hedrick *et al.*, 1969). Under conditions approximating physiological conditions in muscle, phosphorylase *a* and *b* are both predominantly dimeric with molecular weights of 180,000 g/mole. However, the various polymeric forms of phosphorylase are in equilibrium and the equilibrium position can be shifted by the environmental conditions (Metzger *et al.*, 1968; Huang and Graves, 1970; DeVincenzi and Hedrick, 1970). The polymeric state of the protein also affects its enzymatic activity with the dimeric form possessing the greatest catalytic activity.

Phosphorylation of the protein occurs on a serine hydroxyl group. The sequence of the peptide chain around this site is known and its sequence is not analogous to the serine proteases (e.g., trypsin, chymotrypsin) or to phos-

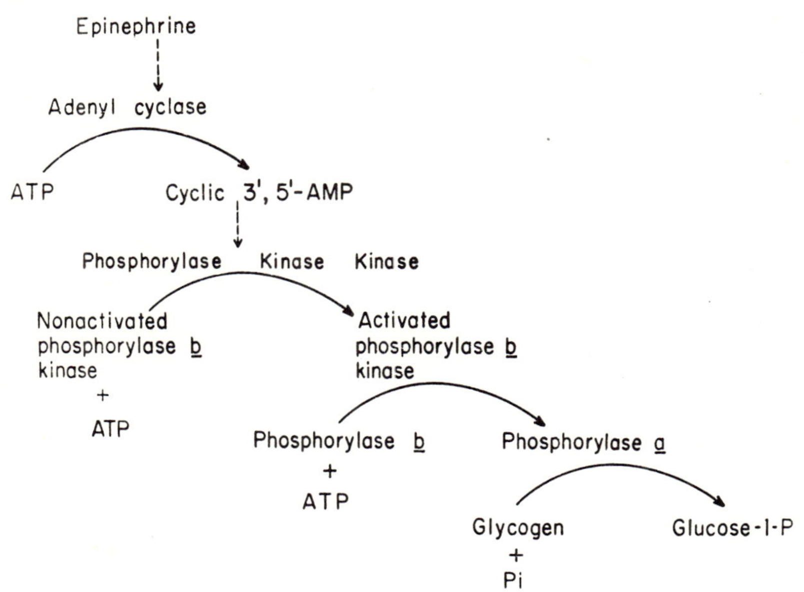

FIG. 1. Scheme for the regulation of phosphorylase activity in muscle (Brostrom *et al.*, 1969).

phoglucomutase (Fischer *et al.*, 1959; Nolan *et al.*, 1964). The stoichiometry of phosphorylation is one serine phosphate/90,000 g protein.

Activation of phosphorylase *b* can also be accomplished by the binding of the allosteric effector AMP. AMP apparently induces activity by altering the protein conformation as does phosphorylation. Phosphorylase *a* activity is stimulated by AMP but is not dependent upon it. As with the serine phosphate sites, the stoichiometry of AMP binding is one AMP binding site/subunit of 90,000 g.

Figure 2 schematically summarizes some of the structural features of the phosphorylase subunit. The enzyme is indeed a complex one, possessing a large number of sites per subunit. The subunit is currently thought to be a single polypeptide chain. The model of phosphorylase is further complicated by the fact that these multiple sites interact with one another, exhibiting both homotropic and heterotropic effects as demonstrated by numerous kinetic and binding experiments.

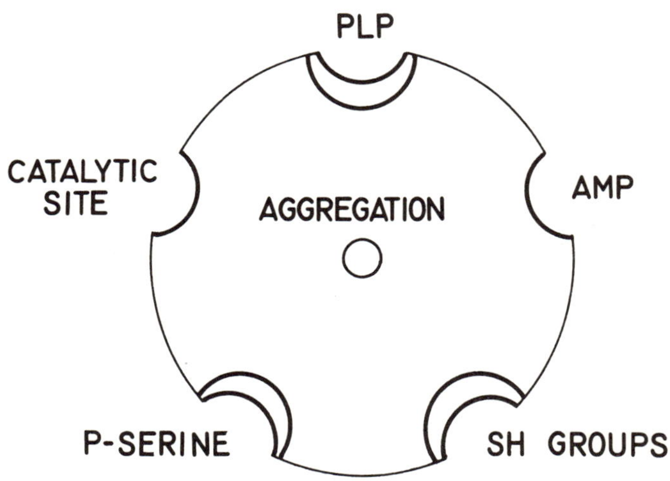

FIG. 2. Schematic representation of sites which affect the structure and function of phosphorylase. The model is based on the phosphorylase *a* monomer, molecular weight of 90,000 g/mole. The catalytic site contains binding sites for phosphate, G-1-P, and glycogen. There are eight sulfhydryl groups per monomer which can be divided into different classes based on the chemical reactivity of the sulfhydryl groups and on the effect that their chemical modification has on structure and catalytic activity. The aggregation site represents at least two different sites since dimers and tetramers of the enzyme can be formed.

III. THE STRUCTURE OF PLP IN PHOSPHORYLASE

The structure of PLP in phosphorylase has been derived from two types of evidence: (1) structural analogies based on the spectral characteristics of model PLP complexes and PLP in phosphorylase, and, (2) the chemical modification of PLP in phosphorylase by $NaBH_4$ reduction and the finding that the vitamin is bound to an ε-amino group of lysine (Kent *et al.*, 1958).

At neutral pH values, phosphorylase exhibits a strong absorption band at 330 nm in addition to the classical protein absorption band at 278nm. A third and minor absorption band can be detected at 415nm. The 330 and 415 nm bands can be interconverted by the addition of acid, base, or conformation-changing agents (e.g., urea or detergents) to the protein. With extremes of pH or denaturation of the protein, the typical spectrum of free PLP is obtained.

The spectral shifts observed in acid or base, and the identification of the ε-amino group of lysine as a link between PLP and the protein moiety have led to the postulated aldamine-Schiff base structures indicated in Fig. 3. The X group was postulated to fulfill the structural requirements dictated by analogy with model compounds in that a substituted aldamine structure absorbs light at 330nm. The nature of the postulated group is unknown, although sulfhydryl, amino, imidazole, or hydroxyl groups could fulfill such a role. However, other PLP structures exist which possess the same absorption characteristics as PLP in phosphorylase. The non-hydrogen-bonded Schiff base in which the pyridoxal ring is in the zwitterionic form or perhaps the carbinolamine derived from it are examples (Christensen, 1958). A chemical argument in favor of the aldamine structure is that at neutral pH, phosphorylase cannot be reduced by $NaBH_4$; therefore, the existence of a Schiff base would seem unlikely. However, an alternative explanation for the lack of $NaBH_4$

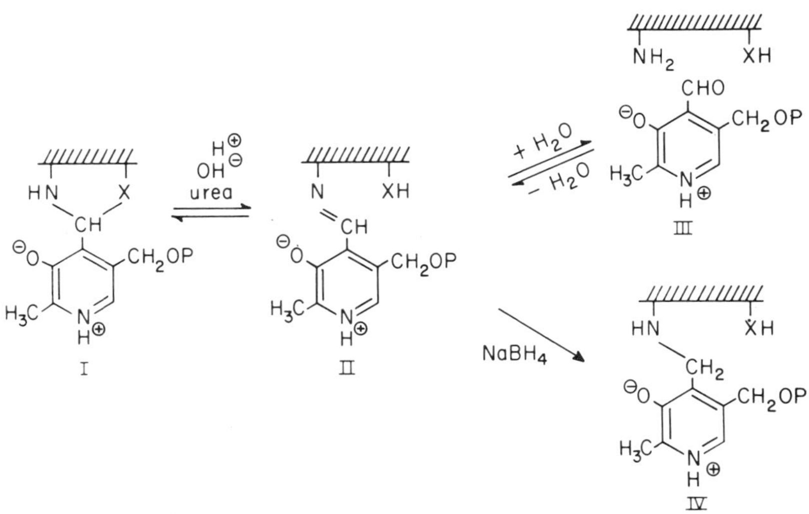

FIG. 3. The structure of pyridoxal 5′-phosphate in phosphorylase (Fischer *et al.*, 1963).

reduction of PLP in phosphorylase at neutral pH is that the Schiff-base form is present, but buried in the protein so that it is inaccessible to $NaBH_4$. As will be discussed later, PLP does appear to be buried in the protein moiety. Thus, even though the structure of PLP is based, to a large extent, on structural analogies with model compounds, the substituted aldamine structure seems a likely one and is consistent with the data, although it is not the only structure that will fit all the observed facts. The identification of the X group must be accomplished in order to establish the correctness of the postulated structure. Recent evidence, although not conclusive, has implicated the sulfhydryl of cysteine as the X group (Kleppe and Damjanovich, 1969; Shaltiel and Zaidenzaig, 1970; Avramovic-Zikic *et al.*, 1970).

Although PLP does not contain an asymmetric center which is optically active, when PLP binds to the phosphorylase protein it is bound in such a way that an asymmetric center is produced. Two lines of evidence support such a conclusion: (1) resolution of the enzyme into PLP and apoenzyme is a stereospecific process as is the reconstitution of the apoenzyme (Shaltiel *et al.*, 1969), and (2) the 330 nm absorbing protein-bound PLP band is optically active as shown by optical rotatory dispersion and circular dichroism measurements (Torchinsky *et al.*, 1965; Johnson and Graves, 1966).

Resolution of phosphorylase involves at least two steps as indicated in Fig. 4. The first step is the modification or deformation of protein structure such that in a second step the PLP can react with a PLP (carbonyl) reagent (Hedrick *et al.*, 1969; Shaltiel *et al.*, 1969). Both steps are remarkably specific, with imidazole-citrate buffer being one of the best deforming agents and L-cysteine one of the most reactive PLP reagents. The PLP reagent is not only structurally specific (analogs such as cysteamine, penicillamine, and homocysteine were either weakly active or not active at all in the resolution process), but also stereospecific as D-cysteine will not cause resolution of the enzyme. A similar situation is obtained in regard to reconstitution of the apoenzyme.

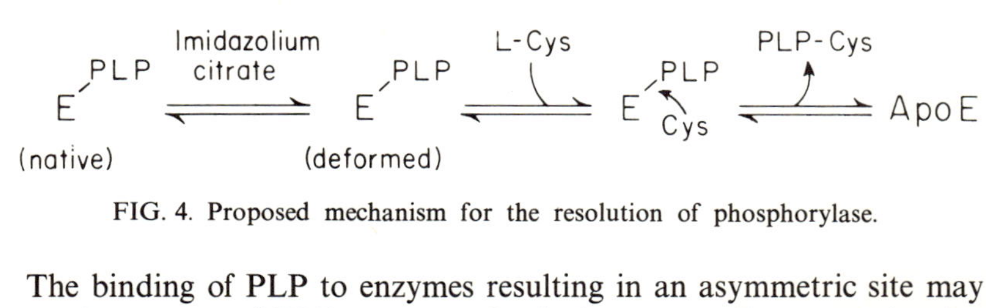

FIG. 4. Proposed mechanism for the resolution of phosphorylase.

The binding of PLP to enzymes resulting in an asymmetric site may be a general property of PLP enzymes. Intrinsic cotton effects associated with PLP absorption bands have been observed for several PLP enzymes in addition to phosphorylase (Torchinsky and Koreneva, 1964; Fasella and Hammes, 1964; Wilson and Meister, 1966). The asymmetry of the site has not been tested by the stereospecificity of apoenzyme and PLP resolution with other PLP enzymes. However, if one assumes an equivalence between the stereospecificity of the resolution process of phosphorylase and stereospecificity of the enzymatic reaction catalyzed by other PLP-containing enzymes, then PLP sites, in general, exhibit stereospecificity. Thus, one can tentatively conclude that vitamin B_6 binding sites are asymmetric as evidenced by the optical properties of the bound PLP absorption bands and the stereospecificity of chemical reactions taking place at that site.

The PLP in phosphorylase would appear to be "buried" or otherwise modified so as to be chemically unavailable for reaction. As mentioned earlier, before PLP can be resolved from the protein moiety, a conformational change induced by deforming agents must take place. Kent (1959) postulated that the phosphorylase PLP spectral shifts in acid or base were caused by conformational changes in the protein, implying an intimate association of the PLP and the protein environment. Attempts at chemically modifying PLP while it is attached to the enzyme, such as substitution of the pyridinium nitrogen, have failed, implying PLP's buried nature. As mentioned previously, the lack of $NaBH_4$ reduction of the PLP-enzyme complex at neutral pH can be interpreted as an inaccessibility of PLP to the reducing agent. In addition, the unusual fluorescence characteristics of PLP indicate that it is not exposed to the solvent (Shaltiel and Fischer, 1967; Cortijo and Shaltiel, 1970). Thus, PLP can be said to be buried in the protein matrix in such a way that an asymmetric site is produced.

The ability to reduce PLP onto an ε-amino group of lysine provides a way of labeling the binding site as first demonstrated by Kent *et al.* (1958). In this regard, the polypeptide sequence around the PLP-lysine has been investigated (Fischer *et al.*, 1963; Forrey, *unpublished observations*).

IV. HYPOTHESES FOR THE ROLE OF PLP IN PHOSPHORYLASE

Many attempts have been made to implicate PLP as functioning in the chemical catalysis of glycogen phosphorolysis. One of the earliest attempts was directed at testing the possibility that G-1-P might displace as a unit the phosphate moiety of PLP (Illingworth *et al.*, 1958). However, since no exchange between the phosphate groups of PLP and G-1-P occurred, this possibility was ruled out. This does not, however, totally eliminate the functioning of the PLP phosphate group in catalysis but only the mechanisms that involve labilization of the phosphate group. The possible functioning of the PLP phosphate moiety in catalysis has recently been raised again, although no conclusive evidence for its involvement can be cited (Kastenschmidt *et al.*, 1968).

One of the most puzzling but most definitive experiments regarding the possible catalytic role of PLP is the $NaBH_4$ reduction of enzyme-bound PLP and its minimal effect on the catalytic properties of the enzyme. $NaBH_4$ reduction decreases the V^{max} to 50–60% of the native enzyme, but otherwise is without effect on its enzymatic characteristics (Fischer *et al.*, 1963). This treatment would completely destroy the catalytic activity of other PLP enzymes. Clearly, the carbonyl group of PLP is not involved in catalysis in the bond-breaking or bond-making sense.

The possible catalytic involvement of the pyridinium nitrogen of PLP has been suggested by Bresler and Firsov (1968). They showed PLP difference spectra of the enzyme in the presence and absence of substrates and attributed these differences to the protonated and unprotonated pyridinium nitrogen forms of PLP. However, as they noted, such spectral changes could be caused by either a direct interaction of PLP with the substrate G-1-P, or by an indirect effect mediated by a conformational change in the protein. Again, unequivocal evidence as to the involvement of this group in catalysis is not available.

Our evaluation of the functional groups on the PLP molecule that have the potential for reacting catalytically with the substrate has been obtained by reconstitution of the apoenzyme with PLP analogs (Shaltiel *et al.*, 1969). This approach was originally used by Illingworth *et al.* (1958) with apophosphorylase *a* and a limited number of analogs. Apophosphorylase *b* was reconstituted with PLP analogs and was then characterized as to its structural (ultracentrifugation) and enzymatic properties. Table 1 illustrates the various analogs tried and their effect on structure and enzymatic activity; Fig. 5 summarizes the results in terms of the positions where limited chemical modification could occur with retention of function. As can be seen, positions 2, 3, and 6 can be modified whereas positions 1, 4, and 5 appear to have specific structural requirements for reconstitution of the enzymatic properties of phosphorylase.

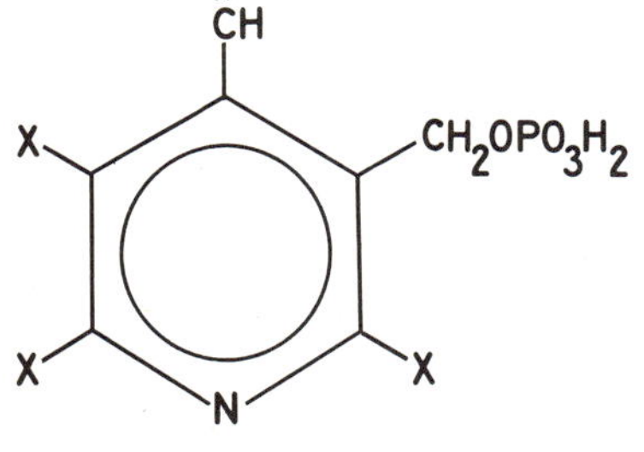

FIG. 5. Schematic representation of the structural analogs used for reconstitution of apophosphorylase *b*. The X refers to positions where limited structural modifications are permitted without loss of PLP function. The pyridinium nitrogen is ring position one with further numbering in a clockwise direction.

The distinction should be made, however, between the ability of the PLP analog to bind and reconstitute the quaternary structure of the enzyme and its ability to regenerate the catalytic properties of the enzyme. A number of analogs possessing the 4'-carbonyl group could reconstitute the structure

TABLE 1. *PLP analogs used to reconstitute apophosphorylase b*

Analog	Ring position modified	Reconstitution: Activity (%)	Reconstitution: Structure (dimeric state)
2-norpyridoxal 5′-phosphate	2	65	+
3-O-methylpyridoxal 5′-phosphate	3	25	+
pyridoxal 5′-phosphate reduced onto phosphorylase	4	60	+
6-methylpyridoxal 5′-phosphate	6	8	
N-methylpyridoxal 5′-phosphate	1	0	−
N-methylpyridoxal	1,5	0	−
ω-methylpyridoxal 5′-phosphate	2	0	
3-hydroxypyridine-4-aldehyde	2,5	0	+
3-O-methylpyridoxal	3,5	0	−
4-deoxypyridoxine 5′-phosphate	4	0	−
pyridoxamine 5′-phosphate	4	0	−
pyridoxamine	4,5	0	−
isopyridoxal 5′-phosphate	4,5	0	
pyridoxine 5′-phosphate	4	0	−
pyridoxine	4,5	0	−
pyridoxic acid 5′-phosphate	4	0	−
pyridoxic acid	4,5	0	−
pyridoxal	5	0	+
pyridoxal 5′sulfate	5	0	+
α^5-pyridoxalacetic acid	5	0	+

of the enzyme but not its catalytic properties. Since the $NaBH_4$-reduced enzyme is enzymatically active, it is concluded that the 4′-carbonyl group functions in binding and in determining the structural aspects of the enzyme, but does not function in a catalytic sense. This then leaves the 5′-phosphohydroxymethyl side chain and the pyridinium nitrogen as potential functional groups in catalysis. It has recently been found that pyridoxal 5′-methylene phosphonate (5′ side chain = $CH_2CH_2PO_3H_2$), in which the ester oxygen of PLP is replaced by a methylene bridge, reactivated apophosphorylase *b* (E. Fischer, *personal communication*). This would seem to weaken the possibility of the 5′-phosphate playing a direct role in catalysis. In addition, most B_6 vitamins are enzymatically active in their phosphorylated form, and it is assumed that the phosphate group is involved in binding of the cofactor. However, closely related analogs such as pyridoxal 5′-sulfate and α^5-pyridoxal acetic acid might be expected to possess some activity in reconstituting the catalytic properties of the enzyme if the 5′-phosphate were only involved in binding. Since they do not, the function of the 5′-phosphate cannot be unequivocably defined in catalytic and/or structural terms. This leaves the pyridinium nitrogen as the only functional group about which no contrary evidence can be cited. As mentioned earlier, it has been implicated as functioning in catalysis by Bresler and Firsov (1968). At the present time, there is no conclusive evidence for or against its functioning in a catalytic sense.

A structural role for PLP in phosphorylase has been amply documented. Table 2 lists the experimental evidence which unequivocably demonstrates that PLP is a determinant of phosphorylase protein structure. When the structural role was originally suggested (Illingworth *et al.*, 1958), it was a much more unique postulate than it would appear to be today. With our current concepts of induced fit, allosterism, conformers, deforming agents, and the dynamic association-disassociation of subunits, the picture of protein structure is one of flexibility or plasticity where environmental conditions or ligands can modulate protein conformation or enzyme activity. Indeed, it is now almost expected that ligand binding by a protein will cause a conformational change of some sort (Koshland and Neet, 1968). The question of the structural role of PLP in phosphorylase is no longer "Does it play a structural role?", but rather, "Is this its predominant functional role?" Some reservations to PLP playing a solely structural role in phosphorylase have been raised based on evolutionary grounds. It might seem that, in terms of cellular metabolism, this is an "expensive way" to maintain a protein in the required conformation by utilizing a nutritionally required vitamin. One might wonder why, after countless eons of evolution, a more conventional way of doing this has not evolved. However, in terms of the cellular metabolism of autotrophs, to synthesize a vitamin might possibly be no more of a "strain" on cellular metabolism than the synthesis of an amino acid. If the requirement for PLP in the functioning of the enzyme was established prior to the evolution of autotrophs to heterotrophs, then it would not necessarily be an "expensive way" to maintain protein conformation, but would only appear to be so. In addition, it might be that phosphorylase at one time possessed enzymatic

TABLE 2. *Evidence confirming the structural role of* PLP *in phosphorylase*

Evidence indicating structural differences between holo- and apoenzyme	Reference
1. dispersity of interacting subunits	Illingworth *et al.*, 1958 Hedrick, 1966 DeVincenzi and Hedrick, 1970
2. optical rotatory dispersion measurements	Hedrick, 1966
3. electrophoretic (disc gel) differences	Hedrick *et al.*, 1969
4. AMP-induced conformational changes, heterotropic interactions lost in apoenzyme	Hedrick, 1966 Kastenschmidt *et al.*, 1968
5. differences in sensitivity to photolysis	Fischer *et al.*, 1963
6. chemical reactivity of sulfhydryl groups	Kleppe and Damjanovich, 1969 Avramovic-Zikic *et al.*, 1970 Shaltiel and Zaidenzaig, 1970
7. PLP spectral shifts of native enzyme	Kent *et al.*, 1958
8. mechanism of resolution of the holoenzyme	Hedrick *et al.*, 1969

activities of the type usually associated with vitamin B_6-containing enzymes (metabolism of amine-containing compounds), but subsequently lost such activity while acquiring glycogenolytic activity. Attempts at finding such vestigial activities have been negative, however (Hedrick and Fischer, 1965). It should indeed be interesting at some future time to compare the structure of various vitamin B_6-containing enzymes to ascertain their evolutionary relations.

The two remaining functions suggested for PLP in phosphorylase—that of controlling enzyme activity and as a PLP depot or reservoir—are somewhat related. They are related in the sense that both would require a dynamic or potentially dynamic equilibrium between enzyme-bound PLP and PLP free in solution. Evidence based on the interchange of free radioactive PLP with enzyme-bound PLP and the necessity to deform the enzyme structure before removing the bound PLP suggests that such an equilibrium is not likely under physiological conditions (Shaltiel *et al.*, Hedrick *et al.*, 1969). Thus it is difficult to envision a control function for PLP in phosphorylase.

A depot or reservoir function of PLP in phosphorylase is based on the observations that in rat skeletal muscle up to 60% of the vitamin B_6 content can be accounted for by that found in phosphorylase, whereas in the case of C_{57} and I strains of mice, 76–96% of muscle vitamin B_6 is present in phosphorylase (for review, see Krebs and Fischer, 1964). Since muscle is the major tissue mass of a mouse or rat, up to one-half of all the vitamin B_6 in the body may be present as PLP in phosphorylase. With this relatively large amount of PLP in phosphorylase, it seems reasonable to postulate a reservoir function for the vitamin in phosphorylase. Under conditions of extreme nutritional stress, where the protein moiety of the enzyme is destroyed, vitamin B_6 could be released to serve other vital functions more necessary to the maintenance of life processes. It has been shown that, in rats and mice maintained on a vitamin B_6-deficient diet, skeletal muscle phosphorylase levels are decreased (Illingworth *et al.*, 1960; Eisenstein, 1962; Lyon and Porter, 1962). However, accurate estimates of the vitamin B_6 content of other body tissues have not been made in order to show that the PLP in phosphorylase is decreasing at a rate faster than other vitamin B_6 sources and thus is potentially serving a reservoir function.

V. FUTURE INVESTIGATIONS AS TO THE ROLE OF PLP IN PHOSPHORYLASE

The only unequivocably proven function of PLP in phosphorylase is its structural role. However, as stated earlier, the question of if this is its only function cannot be answered at present. Knowledge of the amino acid sequence of the protein and its three-dimensional structure would greatly increase our understanding of the structure and role of PLP in phosphorylase. However, it is unlikely that this information will be available for some time to come, as sequence work is still in the very early stages and the effort required to sequence a protein the size of phosphorylase is enormous (E. Fischer, *personal communication*). X-ray diffraction studies have yet to be undertaken on this complex protein.

Thus, we must turn to other, perhaps less definitive, approaches for the near future. Attempts to test the role of PLP in the catalytic sense will require approaches different than those used in the past. To this end, it is suggested that efforts to define the active site of phosphorylase be undertaken with no preoccupation with the potential role of PLP. Attempts at isolating a covalent enzyme-substrate complex analogous to that found in sucrose phosphorylase (Doudoroff *et al.*, 1947; Voet and Abeles, 1970) have not met with success. Yet this does not mean that a covalent enzyme-substrate complex does not exist in the case of phosphorylase. Taking into consideration the chemical nature of phosphorylase substrates, renewed efforts at covalently modifying the catalytic site of the enzyme should be made.

Experiments to estimate the turnover of phosphorylase and of PLP in phosphorylase have not been reported. Such experiments are critical for evaluating the potential role of phosphorylase as a reservoir of vitamin B_6 or the control of phosphorylase activity by PLP. In addition, the role PLP plays in the biosynthesis of phosphorylase and its structural assembly *in vivo* has not been investigated. Thus, the structural role of PLP in phosphorylase has been amply documented, but totally definitive information about its other potential roles has yet to be obtained.

Acknowledgments. The author wishes to express his deep indebtedness to Drs. Edmond H. Fischer and Edwin G. Krebs for introducing him to the complexities of phosphorylase, and to NIH Grants AM-11780 and HD-4906 for financial support.

REFERENCES

Avramovic-Zikic, O., Smillie, L.B., and Madsen, N. B. (1970): The sulfhydryl groups of muscle phosphorylase IV. Reactivities as related to changes in protein structure. *Journal of Biological Chemistry, 245*:1558.

Baranowski, T., Illingworth, B., Brown, D. H., and Cori, C. F. (1957): The isolation of pyridoxal-5-phosphate from crystalline muscle phosphorylase. *Biochimica et Biophysica Acta*, 25:16.

Bresler, S., and Firsov, L. (1968): Spectrophotometric study of the enzyme-substrate complex of phosphorylase *b*. *Journal of Molecular Biology, 35*:131.

Brostrom, M. A., Reimann, E. M., Walsh, D. A., and Krebs, E. G. (1969): A cyclic 3′,5′-AMP-stimulated protein kinase from cardiac muscle. In: *Advances in Enzyme Regulation*, Vol. 8, edited by G. Weber, pp. 191–203. Pergamon Press, New York.

Brown, D. H., and Cori, C. F. (1961): Animal and plant polysaccharide phosphorylases. In:

The Enzymes, Vol. 5, edited by P. D. Boyer, H. Lardy, and K. Myrback, pp. 207–228. Academic Press, New York.

Christensen, H. N. (1958): Three Schiff base types formed by amino acids, peptides, and proteins with pyridoxal and pyridoxal-5-phosphate. *Journal of the American Chemical Society, 80*:99.

Cori, C., and Illingworth, B. (1957): The prosthetic group of phosphorylase. *Proceedings of the National Academy of Sciences (U.S.), 43*:547.

Cortijo, M., and Shaltiel, S. (1970): A built-in fluorescent probe in $NaBH_4$-reduced glycogen phosphorylase. *Biochemical and Biophysical Research Communications, 39*:212.

DeVincenzi, D. L., and Hedrick, J. L. (1967): Reevaluation of the molecular weights of glycogen phosphorylases *a* and *b* using Sephadex gel filtration. *Biochemistry, 6*:3489.

DeVincenzi, D. L., and Hedrick, J. L. (1970): Gel filtration, aggregation, and the enzymatic activity of glycogen phosphorylase. *Biochemistry, 9*:2048.

Doudoroff, M., Barker, H. A., and Hassid, W. Z. (1947): Studies with bacterial sucrose phosphorylase I. The mechanism of action of sucrose phosphorylase as a glucose-transferring enzyme (transglucosidase). *Journal of Biological Chemistry, 168*:725.

Eisenstein, A. B. (1962): The effect of pyridoxine deficiency on liver and muscle phosphorylase. *Biochimica et Biophysica Acta, 58*:244.

Fasella, P., and Hammes, G. G. (1964): An optical rotatory dispersion study of aspartic amino transferase. *Biochemistry, 3*:530.

Fischer, E. H. (1964): Binding of vitamin B_6-coenzymes and labelling of active site of enzymes by sodium borohydride reduction. In: *Structure and Activity of Enzymes,* edited by T. W. Goodwin, J. I. Harris, and B. S. Hartley, pp. 111–120. Academic Press, New York.

Fischer, E. H., Forrey, A. W., Hedrick, J. L., Hughes, R. C., Kent, A. B., and Krebs, E. G. (1963): Pyridoxal 5′-phosphate in the structure and function of phosphorylase. In: *Chemical and Biological Aspects of Pyridoxal Catalysis,* edited by E. E. Snell, P. M. Fasella, A. Braunstein, and A. Rossi-Fanelli, pp. 543–562. Pergamon Press, New York.

Fischer, E. H., Graves, D. J., Crittenden, E. R. S., and Krebs, E. G. (1959): Structure of the site phosphorylated in the phosphorylase *b* to *a* reaction. *Journal of Biological Chemistry, 234*:1698.

Fischer, E. H., and Krebs, E. G. (1966): Relationship of structure to function of muscle phosphorylase. *Federation Proceedings, 25*:1511.

Hedrick, J. L. (1966): An optical rotatory dispersion study of glycogen phosphorylase. *Archives of Biochemistry and Biophysics, 114*:216.

Hedrick, J. L., and Fischer, E. H. (1965): On the role of pyridoxal 5′-phosphate in phosphorylase. I. Absence of classical vitamin B_6-dependent enzymatic activities in muscle glycogen phosphorylase. *Biochemistry, 4*:1337.

Hedrick, J. L., Shaltiel, S., and Fischer, E. H. (1969): Conformational changes and the mechanism of resolution of glycogen phosphorylase *b*. *Biochemistry, 8*:2422.

Hedrick, J. L., Smith, A. J., and Bruening, G. E. (1969): Characterization of the aggregated states of glycogen phosphorylases by gel electrophoresis. *Biochemistry, 8*:4012.

Huang, C. Y., and Graves, D. J. (1970): Correlation between subunit interactions and enzymatic activity of phosphorylase *a*. Method for determining equilibrium constants from initial rate measurements. *Biochemistry, 9*:660.

Illingworth, B., Jansz, H. S., Brown, D. H., and Cori, C. F. (1958): Observations on the functions of pyridoxal 5′-phosphate in phosphorylase. *Proceedings of the National Academy of Sciences (U.S.), 44*:1180.

Illingworth, B., Kornfeld, R., and Brown, D. H. (1960): Phosphorylase and uridinediphosphoglucose-glycogen transferase in pyridoxine deficiency. *Biochimica et Biophysica Acta, 42*:486.

Johnson, G. F., and Graves, D. J. (1966): Circular dichroism and optical rotatory dispersion of glycogen phosphorylase. *Biochemistry, 5*:2906.

Kastenschmidt, L. L., Kastenschmidt, J., and Helmreich, E. (1968): Subunit interactions and their relationship to the allosteric properties of rabbit skeletal muscle phosphorylase *b*. *Biochemistry, 7*:3590.

Kent, A. B. (1959): Studies on muscle phosphorylase. Ph.D. Thesis, University of Washington, Seattle.

Kent, A. B., Krebs, E. G., and Fischer, E. H. (1958): Properties of crystalline phosphorylase *b*. *Journal of Biological Chemistry, 232*:549.

Kleppe, K., and Damjanovich, S. (1969): Studies on the SH groups of phosphorylase *b*. Reaction with 5,5′-dithiobis-(2-nitrobenzoic acid). *Biochimica et Biophysica Acta, 185*:88.

Koshland, D. E., Jr., and Neet, K. E. (1968): The catalytic and regulatory properties of enzymes. *Annual Review of Biochemistry, 37*:359.

Krebs, E. G., and Fischer, E. H. (1964): Phosphorylase and related enzymes of glycogen metabolism. In: *Vitamins and Hormones,* edited by R. S. Harris, I. G. Wool, and J. A. Loraine, pp. 399–410. Academic Press, New York.

Lyon, J. B., Jr., and Porter, J. (1962): The effect of pyridoxine deficiency on muscle and liver phosphorylase of two inbred strains of mice. *Biochimica et Biophysica Acta, 58*:248.

Madsen, N. B., and Cori, C. F. (1956): The interaction of muscle phosphorylase with *p*-chloromercuribenzoate. I. Inhibition of activity and effect on the molecular weight. *Journal of Biological Chemistry, 223*:1055.

Metzger, B. E., Glaser, L., and Helmreich, E. (1968): Purification and properties of frog skeletal muscle phosphorylase. *Biochemistry, 7*:2021.

Nolan, C., Novoa, W. B., Krebs, E. G., and Fischer, E. H. (1964): Further studies on the site phosphorylated in the phosphorylase *b* to *a* reaction. *Biochemistry, 3*:542.

Seery, V. L., Fischer, E. H., and Teller, D. C. (1967): A reinvestigation of the molecular weight of glycogen phosphorylase. *Biochemistry, 6*:3315.

Shaltiel, S., and Fischer, E. H. (1967): Visible fluorescence of glycogen phosphorylase *b*. *Israel Journal of Chemistry, 5*:127p.

Shaltiel, S., Hedrick, J. L., and Fischer, E. H. (1966): On the role of pyridoxal 5′-phosphate in phosphorylase. II. Resolution of rabbit muscle phosphorylase *b*. *Biochemistry, 5*:2108.

Shaltiel, S., Hedrick, J. L., and Fischer, E. H. (1969): Stereospecific requirements for carbonyl reagents in the resolution and reconstitution of phosphorylase *b*. *Biochemistry, 8*:2429.

Shaltiel, S., and Zaidenzaig, Y. (1970): Active site peptides from glycogen phosphorylase. *Biochemical and Biophysical Research Communications, 39*: 1003.

Torchinsky, Y. M., and Koreneva, L. G. (1964): Anomalous rotatory dispersion of aspartate-glutamate transaminase: The effects of carbonyl reagents and substrate analogues. *Biochimica et Biophysica Acta, 79*:426.

Torchinsky, Y. M., Livanova, N. B., and Pikhelgas, V. Y. (1965): The circular dichroism and anomalous rotatory dispersion of muscle phosphorylase *b*. *Biochimica et Biophysica Acta, 110*:619.

Ullmann, A., Goldberg, M. E., Perrin, D., and Monod, J. (1968): On the determination of molecular weight of proteins and protein subunits in the presence of 6M guanidine hydrochloride. *Biochemistry, 7*:261.

Velick, S. F., and Wicks, L. F. (1951): The amino acid composition of phosphorylase. *Journal of Biological Chemistry, 190*:741.

Voet, J. G., and Abeles, R. H. (1970): The mechanism of action of sucrose phosphorylase. Isolation and properties of a *β*-linked covalent glucose-enzyme complex. *Journal of Biological Chemistry, 245*:1020.

Wilson, E. M., and Meister, A. (1966): Optical rotatory dispersion of L-aspartate *β*-decarboxylase and its derivatives. *Biochemistry, 5*:1166.

ADDENDUM

Since submitting this manuscript, several publications have appeared which further define the structure and binding of PLP in phosphorylase.

Johnson, G. F., Tu, J., Bartlett, M. L. and Graves, D. J. (1970): Physical-chemical studies on the pyridoxal phosphate binding site in sodium borohydride-reduced and native phosphorylase. *Journal of Biological Chemistry, 245*:5560–5568.

Jones, D. C., and Cowgill, R. W. (1970): Fluorescence studies on the role of pyridoxal-5′-phosphate in muscle phosphorylase *b. Abstracts of the American Chemical Society Meeting,* Chicago, September 13–18. p. Biol–241.

Shaltiel, S., and Cortijo, M. (1970): The mode of binding of pyridoxal-5′-phosphate in glycogen phosphorylase. *Biochemical and Biophysical Research Communications, 41*:594–600.

Advances in Biochemical Psychopharmacology, Vol. 4.
Raven Press, New York © 1972.

Nature and Roles of Receptor Sites for Amino Acid Transport

Halvor N. Christensen

Department of Biological Chemistry, The University of Michigan, Ann Arbor, Michigan

I. EARLY EVIDENCE FOR PARTICIPATION OF VITAMIN B_6 IN TRANSPORT

It was in the early 1950's that Riggs and I began to think how the amino acid molecule might be grasped during its biological transport. Among a number of nutritional factors we tested, only pyridoxal and pyridoxal 5′-phosphate appeared to stimulate uptake of amino acids by the Ehrlich cell (1963). Ehrlich cells grown in vitamin B_6-deficient mice appeared especially sensitive to this action. The naked protoplast of *Lactobacillus arabinosus* has also been found responsive to pyridoxal, but also at rather high levels (Mora and Snell, 1963). Subsequently Riggs and Walker (1958) found that B_6-deficient rat maintained lowered cellular accumulations of amino acids, relative to the plasma level. In the meantime other investigators found that intestinal tissue rendered deficient in vitamin B_6 or, more specifically, in pyridoxal phosphate in various ways could be stimulated in its amino acid absorption by pyridoxal phosphate (Jacobs and Hillman, 1958; Akedo, Sugawa, Yoshikawa, and Suda,

1960; Jacobs, Coen, and Hillman, 1960; Ueda, Akedo, and Suda, 1960; Braunstein and Vilenkina, 1963).

At the time, Metzler, Ikawa, and Snell (1954) were producing non-enzymatic model reactions for transamination and other catalytic transformations of amino acids, in which Schiff bases were formed between the aldehyde group of pyridoxal and the amino group of the amino acid (1954). Subsequent research has fully confirmed their proposal that similar reactions of the aldehyde group of enzyme-bound pyridoxal phosphate account for the activities of a broad range of vitamin B_6 enzymes.

Our provisional inference was that formation of such Schiff bases might play a key role in the grasping of each amino acid for transport. The simplest supposition was that such a derivative, perhaps stabilized by a metal ion or by hydrogen bonding, might in fact be the molecular species handed across the osmotic barrier of the plasma membrane. As the detailed structural specificity of transport unfolded, however, I became convinced that a three-dimensional, macromolecular recognition site is necessary (1960), and came to doubt that a structure as simple as pyridoxal phosphate could serve alone, even as the "carrier." Such Schiff bases form and dissociate freely in solution; hence no catalytic protein appears necessary to form or to cleave them. Furthermore, if the Schiff-base formation were limited to specific amino acids because of the restrictive effect of a protein component, release of the Schiff base into an aqueous phase would be expected to lead to an exchange of its amino acid component with other amino acids, leading presumably to a loss of structural specificity. By such arguments the hypothetical role of pyridoxal phosphate under examination became, in our estimation, the smaller one of fixing the alpha-amino group to a macromolecule. At the same time, other structures of the macromolecule might be expected to serve, as they do in the B_6 enzymes, to fix the alpha-carboxyl group, the apolar portion of the side chain, and any other functional group of the amino acid substrate.

Some of the interest in pyridoxal phosphate was generated by its ability to form ring compounds by complexing with two neighboring amino groups at once, or with other pairs of groups, for example in cysteine. Some of these reactions are pertinent in other ways to the subject of our symposium, and will no doubt be referred to by others. At one time we hoped that the anomalous and intense transport reactivity of α, γ-diaminobutyric acid and α, β-diaminopropionic acid might be accounted for by such reactions (Christensen, Riggs, Fischer, and Palatine, 1952*a*, *b*). I believe now, however, that we are approaching explanations for the anomalous transport properties of these and related diamino acids along other lines.

II. ACCUMULATING EVIDENCE AGAINST PARTICIPATION OF PYRIDOXAL OR PYRIDOXAL PHOSPHATE

Let me suggest also that nature would be using an overly sophisticated structure if it had selected pyridoxal phosphate to bind the amino group of amino acids to transport receptor sites. Its pyridine nucleus is equipped to destabilize the amino acid molecule in ways that are in fact of no obvious value for transport. We know in this connection that model amino acids designed to be intensely resistant to metabolic modification are transported in entirely typical ways. The substrate shown in Fig. 1, namely 2-aminonorbornyl-2-carboxylic acid (also called 2-aminobicyclo [2, 2, 1] heptyl-2-carboxylic acid, abbreviated in some of our figures BCH), presents a molecule so compact that the transport event can scarcely distort the structure even transiently (Christensen, Handlogten, Lam, Tager, and Zand, 1969). The circumstance that the B_6 coenzyme seems unnecessarily elaborate to us does not of course tell us whether it actually serves or not.

A series of observations that carbonyl reagents in general (semicarbazide, hydroxylamine, isonicotinic hydrazide, aminoxyacetic acid) do not inhibit

FIG. 1. Photograph of atomic model of 2-aminonorbornyl-2-^{14}C-carboxylic acid (2-aminobicyclo [2, 2, 1]-heptane-2-carboxylic acid) to show compactness of the structure. The carboxyl group protrudes at the right, and the amino group above it. Fig. 3 shows a conventional formula for this amino acid.

amino acid transport has been influential in causing us to abandon the idea (Christensen, 1960). Pal and I (1961) also found that the flux relations between labeled amino acids and labeled pyridoxal or pyridoxal phosphate were not such as to support the hypothesis. In the meantime we have come to understand that the transport stimulation by free pyridoxal and pyridoxal phosphate arises indirectly from an inhibition rather than directly from a stimulation of a flux (Christensen, 1963). Furthermore, Heindel and Riggs (1968) have pointed out that subnormal transport in the B_6-deficient rat might arise from decreased growth-hormone release. Tews (1969) found no deficiency in the uptake of AIB or histidine by cerebral cortex slices of B_6-deficient rats.

Nevertheless we are still unable to repudiate the concept absolutely. We now know that a variety of amino acid transport mechanisms exist together, some of which appear to be quite unlike others. Therefore the question has to be asked not just once but for each transport system, and it remains conceivable that a system based on the properties of this versatile coenzyme may eventually be discovered. In the meanwhile we may also speculate if the special sensitivity of intestinal absorption to B_6 could also imply an indirect rather than a direct involvement.

Although we cannot absolutely exclude a participation of vitamin B_6 for every amino acid transport system, in a number of instances we can be fairly sure that it does not play the originally proposed role. My colleagues Piperno and Oxender (1966) have isolated from *E. coli* a binding protein for leucine, isoleucine, and valine that appears to correspond to and be part of a transport system for these three amino acids. Crystals of this protein are shown in Fig. 2. Penrose, Nichoalds, Piperno, and Oxender (1968) have shown that this protein definitely contains no vitamin B_6. Note further that by this time a number of binder proteins corresponding to bacterial transport systems for other amino acids have been isolated (see review by Oxender, 1971), and vitamin B_6 has been reported present in none of them. These binding proteins present just the kind of non-enzymatic receptor sites for which we were looking, but B_6 apparently does not in any known instance serve to give them their exceedingly high binding activity for amino acids. A decade or two ago, as far as I recall, we lacked precedent for such a tight binding of amino acids without the assistance of any prosthetic group.

In the meanwhile my associates and I have continued by means of structure-activity relations to describe the chemical nature of several distinct transport receptor sites in various animal cells. Not only have we failed to see any direct evidence of B_6 involvement, but we have also made the following observations that argue directly that the amino group is probably not grasped by Schiff-base formation.

1. In transport System *A* and in the beta system, one to three α-N-methyl groups can be tolerated on the substrate without complete loss of activity (Christensen, Oxender, Liang, and Vatz, 1965). Schiff-base formation appears to be excluded for such structures.

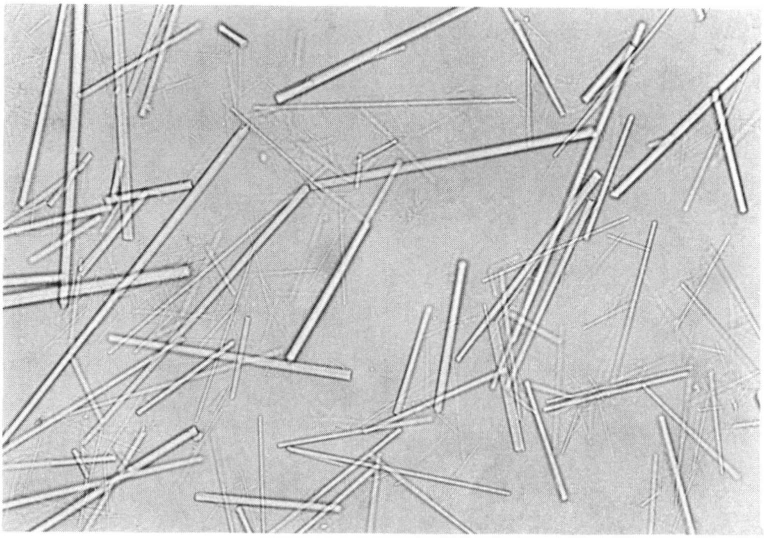

FIG. 2. Crystals of the binding protein for leucine, isoleucine, and valine isolated by Penrose, Nichoalds, Piperno, and Oxender (1968). (Reproduced with permission of the *Journal of Biological Chemistry*.)

2. An increased ease of displacement of a proton from the charged amino group, produced for example by a lowering of its pk_a by a β-fluoro group, appears unfavorable to the transport reactivity of substrates (Christensen and Oxender, 1963). Schiff-base formation does not proceed well without displacement of a proton.

3. Certain polar groups on the sidechain of the amino acid which enhance substrate reactivity have recently been shown to do so not through their lowering of the pK_a of the amino group as we once suspected, but by their enhancing the binding of the cosubstrate Na^+. The position indicated for the Na ion is not such as to suggest its participation in bonding to pyridoxal phosphate (Thomas and Christensen, 1970, 1971).

III. DISCRIMINATION OF TRANSPORT SYSTEMS OF OVERLAPPING FUNCTIONS

By this last item I have abruptly projected my subject into the area of what we have learned in a positive way about the structure of transport receptor sites. Although perhaps I can justify my treating this information in detail here by the negative answer it provides as to a role for vitamin B_6, I think I can justify the subject better because of the intimate relations that can be recognized between amino acid transport and the function of the central nervous system.

Wherever these relations may not be obvious to us, I hope we may be stimulated to discover them in this book.

Table 1 has been prepared to show in a crude way how neutral amino acids tend to divide their transport into cells among several systems. Let me call your attention to the circumstance that typical amino acids divide their transport between two or three different systems. This table was constructed for the Ehrlich cell and should not be taken as closely applicable to every cell or tissue. It is clear, for example, that the overlap among systems is even more intense for intestinal transport, so that the heterogeneity in the absorption of neutral amino acid by this tissue is easily overlooked. As the table illustrates, both Systems *A* and *L* show strong responses to the apolar mass of the sidechain, which, however, needs to be arranged in a linear chain for optimal reaction with System *A*, although the chain can be grossly branched for System *L*. System *A* tolerates polar elements rather well, as does a somewhat similar System ASC. These latter two are Na^+-dependent systems, and I will presently show you how Na^+ assists them in accepting polar structures present in the amino acid sidechain.

The point I want first to emphasize is that only occasionally does a natural amino acid serve as a defining substrate for a given transport system. It is a gross oversimplification to label leucine or phenylalanine as *L*-system substrates, or to use alanine or glycine transport as measures of the function of

TABLE 1. *Approximate survey of reactivities of α-amino acids with several transport systems*

Amino acid	Gly	A	ASC	L
gly, sar	+ + +	+ +	0	±
pro	in some cases	+ + +	±	0
ala, ser	0	+ + +	+ + +	+
thr, homoser	0	+ + +	+ + + +	+ +
α-aminobutyric	0	+ +	+ + +	+ +
$gluNH_2$	0	+ +	+ +	+ +
norval	0	+ + +	+	+ + +
norleu, met	0	+ + + +	0	+ + + +
his, tyr, trn	0	+ +	0	+ +
val	0	+	0	+ +
leu, ileu, phe	0	+	0	+ + + +
Model substrates	*Ly^{+}*	*A*	*ASC*	*L*
MeAIB	0	+ + + +	0	0
BCH	0	0	0	+ + + +
pseudoarg	+ +	0	0	0
DCG	0	0	0	0

Reactivities are compared roughly as observed in the Ehrlich cell and various red blood cells, and do not necessarily correspond in all tissues. A zero implies that no significant reactivity was observed. MeAIB = α-(methylamino)-isobutyric acid; BCH = 2-aminonorbornyl-2-carboxylic acid; pseudoarg = 4-amino-1-guanylpiperidine-4-carboxylic acid; DCG = α, α-dicyclopropylglycine.

System *A*. On the contrary, we have had to make a substantial effort to design substrates typical to those two systems. Literally scores of artificial amino acids had to be investigated before defining substrates could be identified. For System *A*, we added to the alanine molecule a second α-methyl group to make it resistant to metabolism, and an N-methyl group to make it unacceptable to Systems *L* and *ASC*. Although the product, α-(methylamino)-isobutyric acid (MeAIB), serves as a defining substrate (Christensen, Liang, and Archer, 1967), again one has to verify for each new cell and tissue the scope of the system which it defines and the precision with which it defines it. With regard to renal transport, it causes increased urinary losses only of glycine and the prolines (Christensen and Cullen, 1968).

IV. NON-METABOLIZABLE SYSTEM-SPECIFIC MODEL SUBSTRATES FOR TRANSPORT STIMULATE HORMONE RELEASE

In Fig. 1, I have already shown you the structure designed to provide a sufficiently bulky side chain (if I may apply the term side chain to this bicyclic structure) to keep it out of transport systems other than *L*. Table 2 shows what we were trying to accomplish in designing this model substrate. Fundamentally, the earlier idea of branching the side chain (Oxender and Christensen, 1963) was improved on by converting the branches into *bridges*, thereby providing the rigidity already referred to. The remarkable circumstance is that this compound reacts in transport like a branched-chain amino acid such as leucine, not only in animal cells but in yeasts and bacteria. Table 2 also shows some unexpected benefits we got when this bicycloheptyl or norbornyl amino acid was studied. It exists in four isomeric forms, which we can visualize by examining Fig. 3. We were able to show that of these four isomeric forms only one is a fully specific substrate for transport in all cells tested. Only it is taken up at any substantial rate by *E. coli*, and it binds much more strongly

TABLE 2. *Desired features for sidechain of a model substrate for system L that led to the norbornyl amino acid*

1. Bulkiness in 3 dimensions, to prevent reaction with other transport systems
2. Maximal apolarity to increase affinity, but without excessive loss of solubility
3. Absence of α-hydrogen and of chain ends, to produce maximal resistance to metabolic attack

Unsought dividends

1. Existence in 4 isomeric forms with small but decisive differences in distribution of bulk
2. Rigidity of structure, with implications for transport mechanism
3. Action mimicking that of leucine with various biological receptor sites

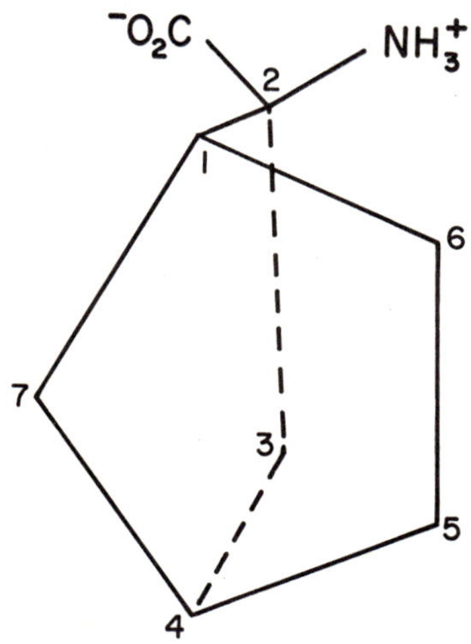

FIG. 3. Representation of the norbornyl amino acid. The isomer shown here corresponds to the biologically active levorotatory form designated $(-)b$. Its dextrorotatory enantiomorph can of course be visualized by viewing the structure as shown in a mirror. The corresponding $(-)a$ and $(+)a$ isomers can be visualized by reversing the pentagon in the foreground so that it points to the right rather than to the left. The discrimination between the corresponding *a* and *b* forms, as by *E. coli* or by the insulin-releasing mechanism of the pancreas, must be based on a fit for the $(-)$ *b* isomer close enough to leave insufficient space for entry of the $(+)$ *a* isomer.

than the other isomers to the leucine-binding protein isolated from that organism. Furthermore, only it shows the remarkable ability to stimulate the release of insulin in the rat and the dog, which I shall describe presently.

The structure actually pictured in Fig. 3 has been found by X-ray crystallography to apply to the isomer with the characteristic biological activity. Figure 4 illustrates the production of hypoglycemia in the rat by this isomer only, an effect arising from the stimulation of insulin secretion (Christensen and Cullen, 1969). Stefan Fajans and his associates of the Laboratory of Endocrinology and Metabolism of our Medical School have shown collaboratively that this model amino acid serves as a close functional analog of leucine in producing this effect. Table 3 lists the properties held in common by the two amino acids (Fajans, Quibrera, Hallam, Pek, Floyd, Christensen, and Conn, 1970). BCH has been shown to be taken up by micro-dissected islets from obese diabetic mice (Christensen, Hellman, Lernmark, Sehlin, Tager and Täljedahl, 1971), only the one isomer causing insulin release.

A detailed approximation of the space available for the amino acid side chain at Site *L* can be reached by placing the α-amino and α-carboxyl groups of this amino acid in fixed positions as in Fig. 3, and then visualizing the four

TABLE. 3 *Features shown in common by leucine and 2-aminonorbornyl-2-carboxylic acid but not by arginine in stimulating insulin release in the dog*

1.	Potentiated by prior treatment with chloropropamide
2.	Inhibited by diazoxide
3.	Not diminished by mannoheptulose
4.	No stimulation of glucagon release

(Fajans, Quibrera, Hallam, Pek, Floyd, Christensen, and Conn, 1970).

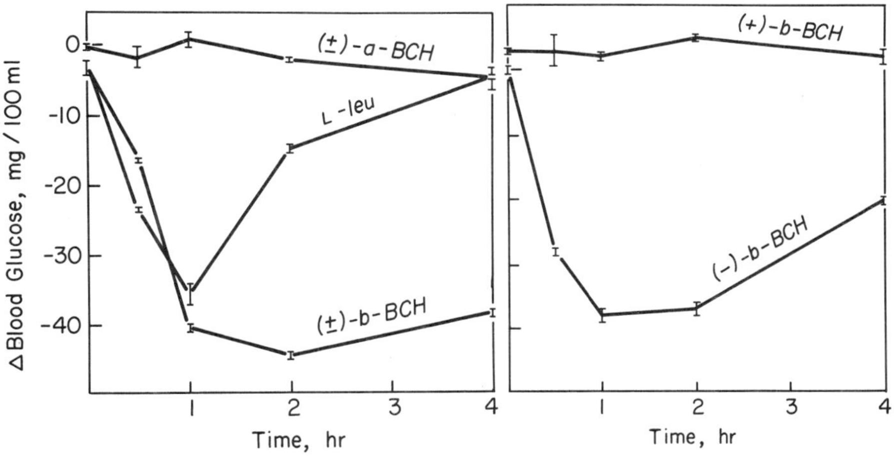

FIG. 4. Comparison of the effect of the isomeric forms of 2-aminonorbornyl-2-carboxylic acid on the blood sugar of the tolbutamide-primed rat. *Left*, comparison of effects of the racemic *a* and *b* isomers and leucine at 8 mmoles per kg body weight. *Right*, comparison of effects of the levorotatory and the dextrorotatory forms of isomer *b* at half the doses used at the left. Procedure as described by Christensen and Cullen (1969). Reproduced with permission from *Biochemical and Biophysical Research Communications*, Tager and Christensen, 1971*a*.

possible dispositions of the side chain. Only the one pictured is accomodated well by the site. The selectivity of *E. coli* for uptake of the four isomers is shown in Fig. 5. The racemic mixture of the *a* isomers shows no uptake, and the dextrorotatory form of the *b* isomer, very slow uptake. Figure 6 shows the strong preference of the hamster intestine for levorotatory *b* over dextrorotatory *b*.

We are now engaged in describing more detailed topography for the *L* site of the Ehrlich cell, and find that a second amino group can be better tolerated on the same side of the cyclohexane ring as the α-amino group, rather than on the opposite side. This amino group apparently is able to release its proton and exist in the uncharged form in the environment occurring at the corresponding position at the receptor site. It is probably a similar event that permits α, γ-diaminobutyric to serve as a substrate for System *A* (Christensen and Liang, 1966). A similar deprotonation would by analogy make it an ideal substrate for System *ASC*, and yet this amino acid does not react with that system. By such methods we gradually expect to describe the microenvironment of all parts of such receptor sites, although I think I should be abusing your interest to go into further detail on such matters in this presentation.

Studies of uptake of amino acids by cortical slices (for reviews see Lajtha, Blasberg, and Levi, 1968, and Lajtha and Ford, 1968) have provided extensive support for the general applicability of the kind of discrimination of transport systems projected in Table 1. We find that the norbornyl amino acid, the

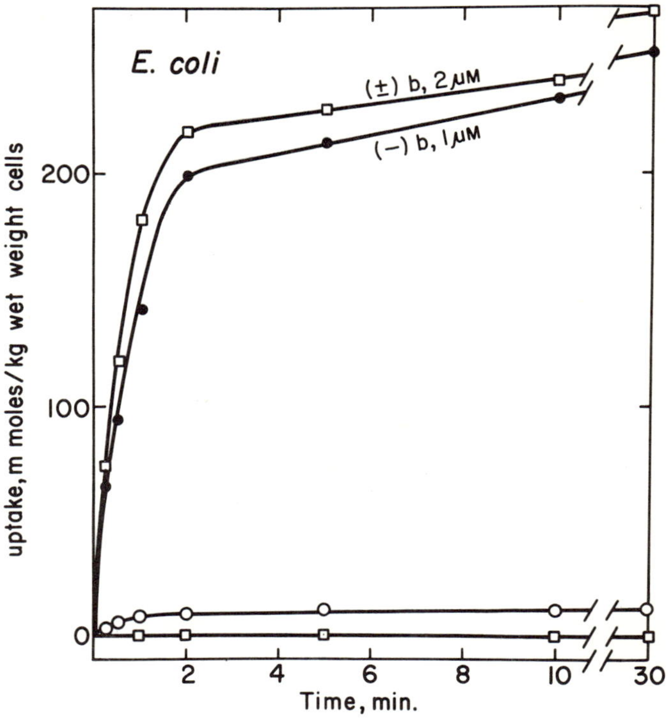

FIG. 5. Specificity of uptake by *E. coli* among the isomers of 2-aminonorbornyl-2-^{14}C-carboxylic acid. No uptake whatever could be observed for the unresolved *a* isomer (lowest line, marked by squares), which means that neither the levo- nor the dextrorotatory form is taken up. Uptake of the unresolved *b* isomer (uppermost curve) approximates very closely the uptake of its levorotatory form (next to top curve), the small uptake of its dextrorotatory antipode (next to lowest curve, marked by open circles), having been shown due to the presence of a small component of the active enantiomorph (Tager and Christensen, 1971*b*).

defining substrate for System *L*, is vigorously taken up by such slices. Unexpectedly, however, the last amino acid mentioned in Table 1, α, α-dicyclopropylglycine, designed by even more extensive crowding in the side chain area to escape specific transport, also gradually enters slices of cerebral cortex (Fig. 7), apparently by the same route as BCH. It also enters the liver in the intact rat (Christensen and Cullen, 1968) and shows a substantial degree of intestinal absorption, which, however, is very easily lost if the physiological function of the intestine is minimally disturbed (Antonioli and Christensen, 1968). Observations of this kind are the basis for our conclusion that the space available at a given type of receptor site for amino acid transport shows a perceptible range from organ to organ and from cell to cell. The unfortunate result is that we need to determine if a given model substrate meets the experimental need presented by each given tissue.

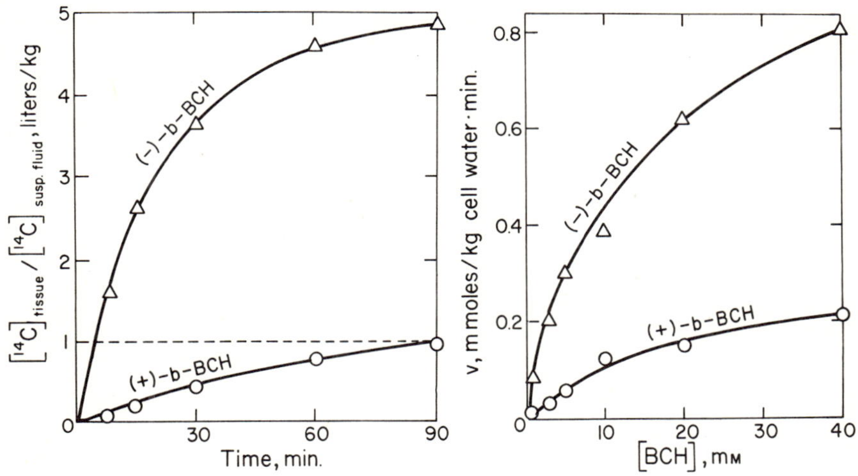

FIG. 6. Comparative uptake by hamster small intestine of the levo- and dextrorotatory forms of the *b* isomer of 2-aminorbornyl-2-^{14}C-carboxylic acid. *Left*, time course of uptake. *Right*, concentration dependence, corresponding to K_m = 14 mM, V_{max} = 1.0 mmoles/kg cell water/min for the levorotatory isomer, and to K_m = 31 mM, V_{max} = 0.37 mmoles/kg cell water/min for the dextrorotatory form. The latter is more uniformly subject to inhibition by dissimilar amino acids (lysine, glycine, etc.) than the former (Tager and Christensen, 1971*b*).

The uptake of amino acids by brain mitochondria shows a vigor not shared with the corresponding organelle of liver and kidney (Navon and Lajtha, 1969), an observation with fascinating possible implications.

We have also designed model substrates for describing the receptor sites

TABLE 4. *Comparison of arginine analogs with arginine as transport substrates and hormone-releasing agents*

	Uptake by Ehrlich cell (K_i, mM)	Plasma hormones in rat	
		Insulin (microunits/ml)	Glucagon (ng/ml)
saline	—	4.1 ± 1.2	0.5 ± 0.3
arginine	0.25	11.7 ± 2.9	1.2
homoarginine	0.06	10.5 ± 0.5	1.3 ± 0.4
1-guanyl-4-aminopiperidine-4-carboxylic acid	2.2	8.5 ± 2.8	1.2 ± 0.2
1-guanyl-DL-4-piperidine glycine	4.0	11.7 ± 3.3	1.3 ± 0.4

Uptake by the Ehrlich cell observed from Krebs-Ringer bicarbonate medium during 1 min at 37°C at a series of concentrations. The amino acids were injected intraperitoneally into fasted rats weighing 100–200 g, 8 mmoles per kg, and blood was taken at 27 min. Insulin and glucagon were measured by double-antibody immuno-assays by Drs. S. Pek and S. S. Fajans, Laboratory for Endocrinology and Metabolism, Department of Medicine, University of Michigan.

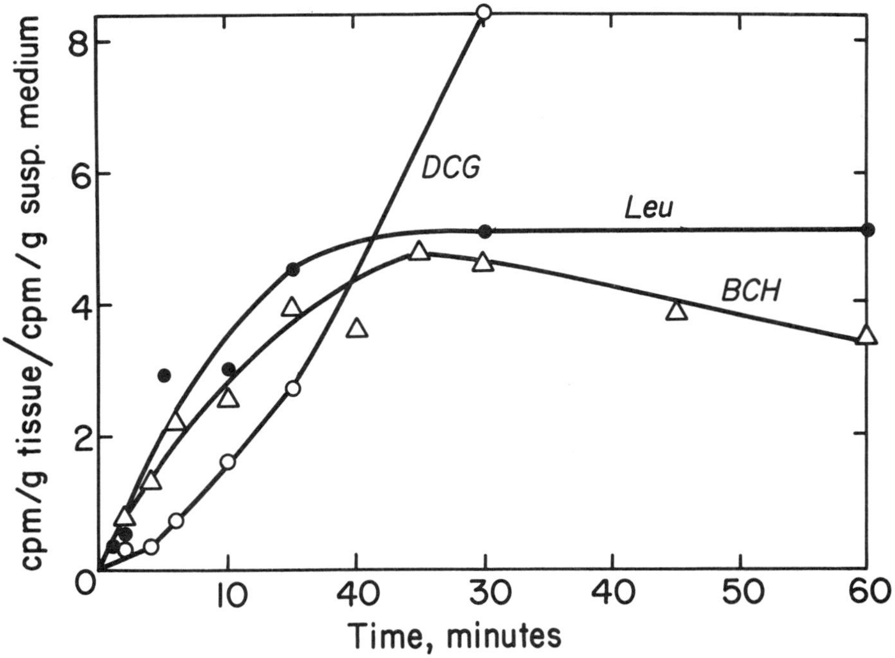

FIG. 7. Comparative time courses of the uptake of leucine-^{14}C, 2-aminonorbornyl-2-^{14}C-carboxylic acid (BCH, 92% isomer *b*), and tritiated α, α-dicyclopropylglycine (DCG) (each 0.2 mM) by slices of cerebral cortex of the rat. Results with isolated cells had led us to hope that the latter amino acid would enter cells only very slowly, and hence could serve to estimate the rate of penetration of all three amino acids into the extracellular space of the slices, so that initial-rate kinetics for cellular uptake might be used. Although an initial lag in its entry into the slices (lower left) encouraged that hope, accumulation of DCG gradually exceeded that of the bicyclic amino acid. Furthermore, the presence of DCG at 24 and 44 mM decreased by 25 and 40%, respectively, the uptake of 0.2 mM BCH during 30 min. Leucine and BCH show mutual inhibitions of each other's uptake at rather lower concentrations. Experiments by Patrick Fleming in the author's laboratory.

for the transport of cationic amino acids. Here the structural features needed were the same as those listed for System *L*(Table 2), except that one more had to be added, namely an unambiguously cationic structure on the side chain, which we obtained with the guanidinium group. In general, diamino acids such as lysine are less than satisfactory because they tend to act both as neutral and as cationic amino acids. Again in this case we found that the specifically active substrates also serve to stimulate hormone release. In this case it is the hormone-releasing action of arginine and lysine, rather than that of leucine, that is mimicked by the arginine analogs shown in Table 4. Arginine and lysine are known to cause the release of both insulin and glucagon, the metabolism of glucose being necessary for these effects. The two non-metabolizable, cyclic arginine analogs pictured in Fig. 8 and described in Table 4 likewise stimulate

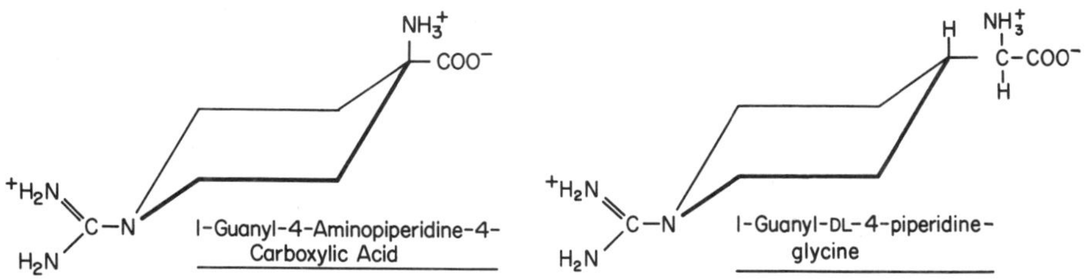

FIG. 8. Model substrates for the cationic amino acid transport system. These were prepared by guanylating the copper salts of 4-aminopiperidine-4-carboxylic acid and DL-4-piperidineglycine. The first of these two precursors was prepared from 4-piperidinol via the N-tosyl derivative in four steps. The compound on the left presents the advantage of having neither a chain end nor an alpha hydrogen, which helps to account for its resistance to metabolism. Furthermore, it presents no stereoisomerism. The compound on the right has an alpha hydrogen and an asymmetric carbon atom, hence it is undoubtedly a DL mixture, with the biological activity largely residing in the L form. Both compounds present an unambiguously cationic structure in the guanidinium group.

the release of both pancreatic hormones in the rat (Christensen, Cullen, and Harrison, 1971).

V. MAY RECEPTOR SITES FOR AMINO ACID TRANSPORT SERVE OTHER ROLES IN THE CENTRAL NERVOUS SYSTEM?

The findings just described call attention to the similarity of receptor sites for transport and those provoking the release of humoral agents. The similarity includes not only structural aspects but also the condition that the active substrate need not be converted to anything else–apparently it needs only to fill a specific receptor site for the responses of transport of hormone release to occur.

Perhaps it would be worthwhile to look for sites that signal for the release of DOPA, in a parallel way. Also, it might be worthwhile to speculate as to whether stimulatory or inhibitory effects are produced in the central nervous system by such amino acids as glutamate, glycine, and γ-amino-butyric acid through their occupation of such receptor sites. It has been proposed that pyridoxal phosphate may participate in the generation of binding sites by which these amino acids produce their effects (Steiner, 1969). I suppose the same possibility can be considered for receptor sites for peptide hormones. But may not these several receptor sites also be similar to or identical with those serving for transport? Heretofore the only related speculations I have seen have to do with the possibility that uptake by cells serves as a mode of *eliminating* (and not of producing) the stimulatory or inhibitory effect of amino acids (*cf.* Kuriyama and Roberts, 1969; Curtis and Johnston, 1970). Is it not conceivable that their filling of the receptor sites is necessary to these humoral effects, and that they cease when completion of transport has again vacated the sites?

The latter speculation gains in plausibility when we consider the migrations of alkali metal ions that are produced by and accompany amino acid transport. In one type of case, the presence of the amino acid causes Na^+ and the amino acid to move in one direction and K^+ in the other, with the result that energy is channeled from the alkali-metal fluxes to amino acid transport (Christensen and Riggs, 1952; Riggs, Walker, and Christensen, 1958–for summaries of recent contributions, see Christensen, 1970, and Schultz and Curran, 1970). This behavior implies that certain amino acids are able to convert a binding site for K^+ into a binding site for Na^+, an effect that is much needed to account for specific alkali-metal transport. It may be recalled that over two decades ago Stern, Eggleston, Hems, and Krebs (1949) showed that the presence of glutamate stimulates the retention of K^+ by brain slices, although as far as I know no flux linkage has been established between the two. This result may mean that glutamate transport assists in the development of alkali-metal gradients. At the same time, the property of Na^+ dependence is also well known for transport of amino acids into brain (*cf.* Lajtha and Ford, 1968).

In a quite different situation, only balanced exchanges of an amino acid plus Na^+ from one compartment for an amino acid plus Na^+ from the other compartment have been observed during amino acid transport, with as many as five Na ions migrating for each amino acid molecule. In that case the amino acid apparently does not have to be displaced from the ternary complex for an exchange diffusion of Na^+ to be set up (Koser and Christensen, 1968; Thomas and Christensen, 1971). It seems to me from such examples that one could reasonably suggest that entry of an amino acid into a transport receptor site might lead to depolarization of the membrane. It also seems conceivable to me that entry of an amino acid into another type of transport receptor site might cause hyperpolarization of the membrane. It remains to be determined if transport systems with the necessary properties actually function in the neurone.

VI. EVIDENCE FOR JUXTAPOSITION AND BONDING BETWEEN AMINO ACIDS AND NA^+ WHEN THEY COMBINE AT TRANSPORT RECEPTOR SITES.

Assuming that by these suggestions I have justified your interest in Na^+-dependent amino acid transport, I want to turn to consideration of the system designated *ASC* in Table 1. *ASC* stands for *alanine*, *serine*, and *cysteine*; the abbreviation serves to remind us that these amino acids and their higher linear homologs are the typical substrates. Table 5 provides a summary description of it. The system has been described in most detail for the rabbit reticulocyte

TABLE 5. *Description of System ASC for neutral amino acid transport*

Formally defined for: The Ehrlich cell

Substrates: Characteristic, 3-and 4-carbon aliphatic and hydroxyaliphatic amino acids, proline, and cysteine. No inhibition detected by MeAIB, MeAla (discrimination from *A* system).

V_{max}: Characteristically variable; several amino acids inhibit without migrating detectably by it.

Dependency on Na^+: First order.

pH sensitivity: Intermediate. Characteristically less than for A system.

Stereospecificity: Exceptional

Exchanging properties: Weak

Other occurrences: In rabbit reticulocyte, pigeon erythrocyte, but with strong exchanging action, and with several sodium ions migrating for each amino acid molecule. Ubiquitous?

and the pigeon red blood cell (Christensen, Liang, and Archer, 1967; Eavenson and Christensen, 1967). It is in these cells that the highest coupling ratios between Na^+ flux and amino acid flux have been seen. We have recently been able to obtain evidence of just where the Na ion binds at the receptor site in relation to the amino acid cosubstrate. I should like to spend the rest of my time tracing that argument.

It began with the discovery that the transport of cationic amino acids (carried of course by an entirely different system) can be inhibited not only by another cationic amino acid, but also by the combination of a suitable neutral amino acid plus Na^+. Furthermore, exchanges occur across the membrane between the cationic amino acid, on the one hand, and between the neutral amino acid and Na^+ on the other. We were led to see that in producing this cooperative action the Na ion probably occupies the position otherwise taken by the cationic structure of the side chain of the basic amino acid (Christensen and Antonioli, 1969; Christensen, 1969; Christensen and Handlogten, 1969; Christensen, Handlogten, and Thomas, 1969). We found that certain groups on the side chain of the amino acid (notably the hydroxyl or mercapto group) serve to sense the position occupied by the alkali-metal ion. First of all, a properly positioned hydroxyl group greatly increases the cooperative action between the two substrates. Secondly, as Fig. 9 illustrates, a sharp change in the selectivity among the alkali-metal ions occurs when a hydroxyl group is present on one specific carbon atom. In the rabbit reticulocyte (*top*) Na^+ and K^+ become far more effective than the other cations when the hydroxyl is on carbon 4. In the Ehrlich cell (*bottom*) a striking inversion in the preference between Li^+ and Na^+ takes place, in favor of Na^+. But for this cell the change occurs when the hydroxyl group is on carbon 5. We take these changes to suggest that the alkali metal probably forms a bridge between the receptor site and the oxygen of the amino acid side chain, provided that the

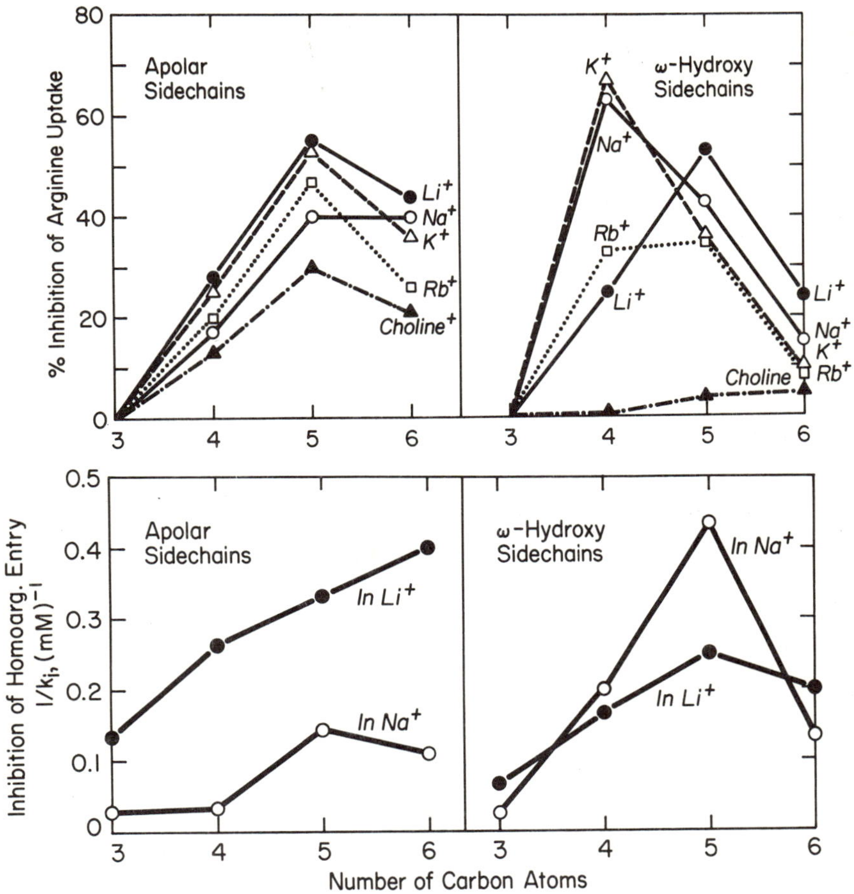

FIG. 9. Cooperative inhibition of cationic amino acid transport by a neutral amino acid plus an alkali-metal cation. Effect of a hydroxyl group on carbon 4 (in the rabbit reticulocyte, top) or on carbon 5 (in the Ehrlich cell, bottom) on the selectivity among the alkali metals. *Top*: In the rabbit reticulocyte, when the hydroxyl group is on carbon 4, the reactivity of Na^+ and K^+ increases abruptly, relative to other alkali metals.
Bottom: For the Ehrlich cell, when the hydroxyl group is on carbon 5, the reactivity of Na^+ increases sharply relative to that of Li^+. Much weaker inhibition of the uptake of the test cationic amino acid is seen when the cation is K^+, Rb^+, or Cs^+, decreasing in that order. In each case the results in the left part of the figure are for amino acids with apolar sidechains, whereas their omega-hydroxyl derivatives appear to the right. (Reproduced, with permission, from the *Journal of Biological Chemistry*; Thomas, Shao, and Christensen, 1971.)

oxygen is presented at a suitable position so that the distance at that receptor site is correct for the preferred alkali-metal ion (Thomas, Shao, and Christensen, 1971).

We were then led to see that the same kind of an association is undoubtedly established between the two cosubstrates in System *ASC* (Christensen, Thomas, and Handlogten, 1969; Thomas and Christensen, 1970, 1971). In this case neutral amino acids and Na^+ are the cosubstrates, and cationic amino acids, e.g., arginine, serve unexpectedly as inhibitors. Again the position of the hydroxyl group on the neutral amino acid is decisive in establishing a cooperativity between it and Na^+, although the position is perceptibly different from that for System *Ly⁺*. In System *ASC* only Na^+ can serve as cosubstrate, although Li^+ functions as an inhibitor competitive with Na^+.

One other signal is seen in System *ASC* when the hydroxyl group is optimally positioned (Table 6). The coupling ratio, namely the ratio of the amino acid-dependent Na^+ flux to the Na^+-dependent amino acid flux, rises sharply, compared with the value seen in the absence of the side chain hydroxyl group. We interpret this result to mean that the hydroxyl group makes it easy for Na^+ to be dissociated from the ternary complex, without any need for the amino acid to be released. The sulfhydryl group functions in a similar way, the coupling ratio observed with cysteine being the highest of all (Koser and Christensen, 1971). Such effects could ultimately mean that entry of a single amino acid molecule into a receptor site might occasion the migration across the membrane of many alkali-metal ions, thus greatly changing the physiological significance of the event. It would be very interesting if transport sites with such properties occur in the synaptic area.

Table 7 summarizes the points of argument by which we are led to our conclusion as to the geometric relations between the two cosubstrates at Site *ASC*. Only point 6 in this list has not yet been documented. Table 8 shows that the hydroxyl group of proline (whether on carbon 3 or carbon 4) must be oriented *trans* to the carboxyl group if the substrate action of *proline* is to be enhanced and if cooperativity with Na^+ is to occur (Thomas and Christensen, 1970). This is perhaps the clearest evidence of all regarding the position taken by Na^+ at the receptor site. It indicates that the direction in which the hydroxyl group projects is critical.

TABLE 6. *Effect of a hydroxyl group on the coupling ratio, $\Delta V_{Na^+}/\Delta V_{a\cdot a\cdot}$, for entry by System ASC of pigeon RBC*

Unhydroxylated Amino Acid		Hydroxylated Amino Acid	
proline	0.2	4-hydroxyproline	3.2
alanine	2.5	serine	4.0
α-amino-*n*-butyric	(?)	cysteine	5.0
		threonine	4.5

(Koser and Christensen, 1971).

The uptake of ^{14}C-labeled amino acid from tris-(hydroxymethyl) aminomethane buffered medium was measured in the presence and absence of Na^+, substituting choline isoosmotically for Na^+. Simultaneously the uptake of $^{22}Na^+$ measured in the presence and absence of the amino acids. Time intervals were typically 4 min, although longer intervals gave similar ratios. Various concentrations of amino acids and Na^+ gave similar coupling ratios.

TABLE 7. *Evidence for nature of interaction between Na^+ and neutral amino acids in systems ASC and Ly^+*

1. Competition between (neutral a.a. + Na^+) and (cationic a.a. without Na^+).
2. Guanidine and methylguanidine cations are more reactive than NH^+_4 in association with neutral a.a., in correspondence to the greater reactivity of arginine than lysine.
3. Hydroxyl group on a specific carbon atom causes mutual enhancement of transport reactivity of the neutral a.a. and Na^+.
4. When hydroxyl group is on the right carbon atom selectivity among alkali-metal ions is sharply modified.
5. Coupling ratio, Na^+ to neutral a.a., is sharply increased in *ASC* of pigeon RBC when the hydroxyl group is on the right carbon atom.
6. The hydroxyl group of hydroxyproline must be *trans* to the carboxyl group.

Figure 10 represents in a diagrammatic way the geometric arrangements we visualize. This diagram shows four components of the receptor site as sources

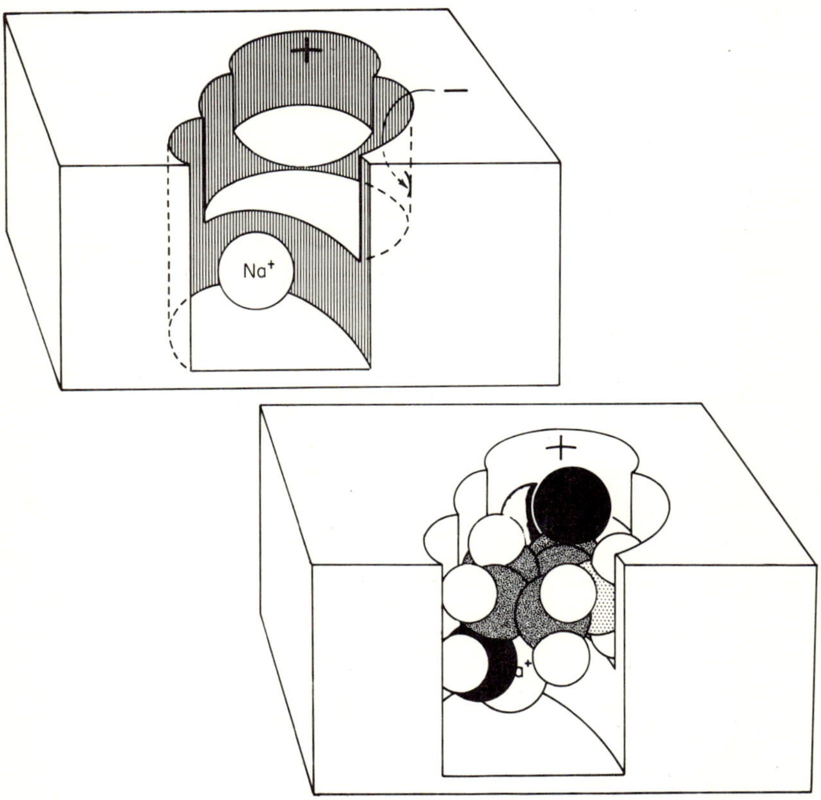

FIG. 10. Diagrammatic visualization for transport site *ASC*, showing Na^+ (upper sketch) and both Na^+ and hydroxyproline (lower sketch) in the proposed relative positions. It is not implied that Na^+ needs to enter the site first; indeed, it appears that Na^+ can be transported without the necessity that the amino acid be released. (Reproduced with permission from *Biochemical and Biophysical Research Communications*; Thomas and Christensen, 1970.)

of binding: (a) a recognition site for the carboxyl group; (b) a recognition site for the amino or imino group (note that this would be the place for the carbonyl group of pyridoxal phosphate to take hold, if indeed it had that function); (c) considerable areas that offer apolar bonding to hydrophobic aspects of the side chain; and (d) a structure recognizing the hydroxyl or sulfhydryl group, if one is present. This structure we believe, on the evidence already summarized, to be the Na ion, properly positioned by bonding groups of the underlying macromolecular structure.

Note in Table 8 that introduction of a 3-hydroxyl group in the *cis* orientation is not a neutral factor but instead is distinctly unfavorable to the ability of proline to serve as a substrate. This result points to the likelihood that apolar bonding occurs with the *cis* aspect of 3-methylene group of the substrate. We can refer to the diagram of Fig. 10 to visualize this function of the methylene group represented by carbon 3. We had already anticipated that this carbon atom is involved in apolar bonding to the site, because alanine is a vastly

TABLE 8. *Effect of a side-chain hydroxyl group on the reactivity of amino acids with transport system ASC*

	Pigeon RBC				Rabbit reticulocyte
Amino or imino acid	K_i at $[Na^+]_0$ = 25 mN (mM)	K_i at $[Na^+]_0$ = 118 mN (mM)	K_i^{25}/K_i^{118}	Stim. of ala exodus (%)	Stim. of ala exodus (%)
proline	2.0	1.8	1.1	16	8
DL-3-*cis* hydroxyproline	11	11	1.0	0.6	—
DL-3-*trans* hydroxyproline	0.08	0.07	1.1	15	—
L-4-*cis* hydroxyproline	2.0	2.0	1.0	4	2
L-4-*trans* hydroxyproline	0.27	0.16	1.7	31	13
L-*threo* threonine	0.22	0.085	2.6	24	—
DL-*threo* threonine	0.27	0.10	2.7	23	—
DL-*allo* threonine	0.24	0.10	2.4	22	—

These results show three indications of reactivity: (1) the values *per se* of K_i (columns 2 and 3) which at 25 mN Na^+ especially receives a high component from the affinity with which the amino acid binds; (2) In column 4, the response of K_i to elevation of $[Na^+]$ from 25 to 118 mN, as measured by the ratio of the K_i value at 25 mN Na^+ to that at 118 mN Na^+; and (3), in the two columns at the right, the extent to which the test amino acid stimulates the exodus of alanine. These second and third measures represent the ability of the amino acid to serve as a substrate rather than merely to bind at the site. K_i values were observed for inhibition of uptake of ^{14}C-serine during 3-min intervals. The K_i values for DL-amino acids have been halved to provide approximate values for the L component. For the two columns at right, red blood cells were first brought to about 1 mM in ^{14}C-alanine by a 30-min incubation. Decline of ^{14}C content of the cells was then observed during 10 min (or 5 min for threonine) at 37°C at 20 mM external levels of the amino acid listed and at $[Na^+]$ = 118 mN (Thomas and Christensen, 1970).

better substrate than glycine. Furthermore, L-alanine and Na^+ show strong cooperation in reacting with the system; hence it appears likely that apolar bonding helps to prepare the recognition site for Na^+, since the methyl group of alanine does not seem at all likely to participate in any way in direct bonding to the alkali-metal ion. Accordingly, we believe that part of the interaction between the two cosubstrates is mediated through the structure of the macromolecular site. Given a favorably positioned polar group on the amino acid side chain, we conclude that another part of the interaction arises from direct linkage between the two bound substrates. For example, if the substrate is serine rather than alanine, the Na ion fixes the hydroxyl group in the *trans* orientation with the result that apolar bonding can still occur with the *cis* aspect of the carbon-3-methylene group. At the same time the reaction with Na^+ is strengthened. The same responses presumably occur to the 3-hydroxyl group of threonine (Table 8), whether it is in the *threo* or *allo* form, because in either case the hydroxyl group can take the *trans* orientation. Only when the atoms are constrained by a ring structure as in the prolines are we able to recognize the necessity that the hydroxyl group must be able to take the *trans* position. It is probably only one aspect of the role of Na^+ in System *ASC* that it serves to draw the polar group out of a position that would otherwise cause interruption of essential apolar aspects of the substrate. This function of Na^+ seems likely to be equally important in the positioning of key hydroxyl groups of monosaccharides by Na^+-dependent systems for transport (Christensen, 1970).

VII. CONCLUDING COMMENT

An extensive comprehension of amino acid transport is likely to prove necessary for an adequate understanding of the function of the central nervous system. Disturbances in the economy of vitamin B_6 cause disturbances in the distribution of a number of important amino acids; disturbances of their transport inevitably also change their distribution and must therefore also be taken into account. Here I offer in conclusion a suggestive and preliminary list of ways in which the transport step in the handling of amino acids may play important roles in central nervous function.

1. The balanced amino acid nutrition of the developing central nervous system almost certainly will suffer if any highly transport-reactive amino acid is maintained at excessively high values, e.g., the high phenylalanine levels of phenylketonuria (McKean and Boggs, 1966; McKean, Boggs, and Peterson, 1968; McKean and DeRafols, 1966). I have urged attention to this factor for a number of years (Christensen, 1953).

2. The occupation of receptor sites may stimulate release of humoral factors (in addition to insulin and glucagon, perhaps also growth hormone or DOPA).

3. Amino acid transport systems will produce the release and removal of amino acids that have functional inhibitory or stimulatory actions in the brain or spinal cord.

4. Amino acid transport systems may condition the distribution of the alkali-metal ions; indeed occupation of transport sites could conceivably be the means by which certain amino acids exert central stimulatory or inhibitory effects.

Acknowledgment. Work cited here was supported in part by grants from the National Cancer Institute and, in recent years, from the Institute for Child Health and Human Development (HD-1233), National Institutes of Health, U.S.P.H.S.

REFERENCES

Akedo, H., Sugawa, T., Yoshikawa, S., and Suda, M. (1960): Intestinal absorption of amino acids. I. Effect of vitamin B_6 on the absorption of L-amino acids through the intestine. *Journal of Biochemistry (Tokyo), 47*:124.

Antonioli, J. A., and Christensen, H. N. (1968): A mode of intestinal absorption unusually dependent on physiological state. *American Journal of Physiology, 215*:*951*.

Christensen, H. N., and Riggs, T. R. (1952): Concentrative uptake of amino acids by the Ehrlich mouse ascites carcinoma cell. *Journal of Biological Chemistry, 194*:57.

Christensen, H. N., Riggs, T. R., Fischer, H., and Palatine, I. M. (1952*a*): Amino acid concentration by a free cell neoplasm: Relations among amino acids. *Journal of Biological Chemistry, 198*:1.

Christensen, H. N., Riggs, T. R., Fischer, H., and Palatine, I. M. (1952b): Intense concentration of α, γ-diaminobutyric acid by cells. *Journal of Biological Chemistry, 198*:17.

Christensen, H. N. (1953): Metabolism of amino acids and proteins. In: *Annual Review of Biochemistry,* 22:233–260. (see p. 235)

Christensen, H. N. (1960): Reactive sites and biological transport. *Advances in Protein Chemistry, 15*:239.

Christensen, H. N. (1963): The relation of vitamin B_6 to transport. In: *Proceedings of the Symposium on Chemical and Biological Aspects of Pyridoxal Catalysis.* p. 533. Pergamon Press, Oxford.

Christensen, H. N., and Oxender, D. L. (1963): The acid strength of the amino group as a factor in the transport of amino acids. *Biochimica et Biophysica Acta, 74*:386.

Christensen, H. N., Oxender, D. L., Liang, M., and Vatz, K. A. (1965): The use of N-methylation to direct the route of mediated transport of amino acids. *Journal of Biological Chemistry, 240*:3609.

Christensen, H. N., and Liang, M. (1966): Transport of diamino acids into the Ehrlich cell. *Journal of Biological Chemistry, 241*:5542.

Christensen, H. N., Liang, M., and Archer, E. G. (1967): A distinct Na^+-requiring transport system for alanine, serine, cysteine and similar amino acids. *Journal of Biological Chemistry, 242*:5237.

Christensen, H. N., and Cullen, A. M. (1968): Effects of nonmetabolizable analogs on the distribution of amino acids in the rat. *Biochimica et Biophysica Acta, 150*:237.

Christensen, H. N. (1969): Some special kinetic problems of transport. In: *Advances in Enzymology and Related Areas of Molecular Biology,* vol. 32. Edited by F. F. Nord. p. 1. Wiley-Interscience, New York.

Christensen, H. N., and Antonioli, J. A. (1969): Cationic amino acid transport by the rabbit reticulocyte. Na^+-dependent inhibition of Na^+-independent transport. *Journal of Biological Chemistry, 244*:1497.

Christensen, H. N., and Cullen, A. M. (1969): Behavior in the rat of a transport-specific, bicyclic amino acid. *Journal of Biological Chemistry, 244*: 1521.

Christensen, H. N., and Handlogten, M. E. (1969): Reactions of neutral amino acids plus Na^+ with a cationic amino acid transport system. *FEBS Letters, 3*: 14.

Christensen, H. N., Handlogten, M. E., and Thomas, E. L. (1969): Na^+-facilitated reactions of neutral amino acids with a cationic amino acid transport system. *Proceedings of the National Academy of Sciences, 63*: 948.

Christensen, H. N., Handlogten, M. E., Lam, I., Tager, H. S., and Zand, R. (1969): A bicyclic amino acid to improve discriminations among transport systems. *Journal of Biological Chemistry, 244*: 1510.

Christensen, H. N., Thomas, E. L., and Handlogten, M. E. (1969): Features of amino acid structure enhancing or obstructing cosubstrate reactivity of Na^+ in transport. *Biochimica et Biophysica Acta, 193*: 228.

Christensen, H. N. (1970*a*): Introduction to Symposium on Membranes and Transport. *FEBS Symposium, 20*: 81.

Christensen, H. N. (1970*b*): Linked ion and amino acid transport. In: *Membranes and Ion Transport,* Vol. 1. edited by E. E. Bittar. p. 365. Wiley-Interscience, London.

Christensen, H. N., Cullen, A. M., and Harrison, L. I. (1971): Reciprocal transport interactions between pancreatic hormones and non-metabolizable model amino acids. *Federation Proceedings: 30* : 200 abs.

Curtis, D. R., and Johnston, G. A. R. (1970): Amino acid transmitters. In: *Handbook of Neurochemistry,* Vol. 4. edited by A. Lajtha. p. 115. Plenum Press, New York.

Eavenson, E., and Christensen, H. N. (1967): Transport systems for neutral amino acids in the pigeon erythrocyte. *Journal of Biological Chemistry, 242*: 5386.

Fajans, S. S., Quibrera, R., Hallam, G. L., Pek, S., Floyd, J. C., Jr., Christensen, H. N., and Conn, J. W. (1970): Stimulation of insulin release by an amino acid (AA) which is unmetabolizable. *Clinical Research, 18*: 538.

Heindel, J. J., and Riggs, T. R. (1968): Growth hormone stimulation of amino acid transport in vitamin B_6 deficient rats. *Federation Proceedings, 27*: 553.

Hellman, B., Lernmark, Å., Sehlin, J., Tager, H. S., and Täljedahl, I. B. (1971): *In Vitro* Stimulation of insulin release by nonmetabolizable, transport-specific amino acids. *Biochimica et Biophysica Acta, 241* : 341.

Jacobs, F. A., and Hillman, R. S. L. (1958): Intestinal absorption of the methionine isomers as affected by deoxypyridoxine. *Journal of Biological Chemistry, 232*: 445.

Jacobs, F. A., Coen, L. J., and Hillman, R. S. L. (1960): Influence of pyridoxine, pyridoxal phosphate, deoxypyridoxine and 2,4-dinitrophenol on methionine absorption. *Journal of Biological Chemistry, 235*: 1372.

Koser, B., and Christensen, H. N. (1971): Effect of substrate structure on coupling ratio for Na^+-dependent transport of amino acids. *Biochimica et Biophysica Acta, 241: 9.*

Kuriyama, K., and Roberts, E., (1969): Association of the γ-aminobutyric acid system with subcellular particulates in chick and mouse brain. *Annals of the New York Academy of Science, 166*: 326.

Lajtha, A., Blasberg, R., and Levi, G. (1968): Control of cerebral amino acid concentrations. In: *Protein Nutrition and Free Amino Acid Patterns,* edited by J. H. Leathem, New Brunswick, 187-205.

Lajtha, A., and Ford, D. H., editors (1968): *Brain-Barrier Systems. Progress in Brain Research,* vol. 29. Elsevier, Amsterdam.

McKean, C. M., and Boggs, D. E. (1966): Influence of high concentrations of phenylalanine on the amino acids of cerebrospinal fluid and blood. *Proceedings of the Society for Experimental Biology and Medicine, 122*: 987.

McKean, C. M., Boggs, D. E., and Peterson, N. A., (1968): The influence of high phenylalanine and tyrosine on the concentrations of essential amino acids in brain. *Journal of Neurochemistry, 15*:235.

McKean, C. M., and DeRafols, W. (1966): The effect of high circulating phenylalanine on the amino acids in blood and cerebrospinal fluid. *Federation Proceedings, 25*:364.

Metzler, D. E., Ikawa, M., and Snell, E. E. (1954): A general mechanism for vitamin B_6-catalyzed reactions. *Journal of the American Chemical Society, 76*:648.

Mora, J., and Snell, E. E., (1963): The uptake of amino acids by cells and protoplasts of *S. faecalis. Biochemistry, 2*:136.

Navon, S., and Lajtha, A. (1969): The uptake of amino acids by particulate fractions from brain. *Biochimica et Biophysica Acta, 173*:518.

Oxender, D. L., and Christensen, H. N. (1963): Distinct mediating systems for the transport of neutral amino acids by the Ehrlich cell. *Journal of Biological Chemistry, 238*:3686.

Oxender, D. L. (1971): Amino acid transport in microorganisms. In: *Metabolic Pathways. Metabolic Transport*, Vol. IV, edited by L. E. Hokin. Academic Press, New York (*in press*).

Pal, P. R., and Christensen, H. N. (1961): Uptake of pyridoxal and pyridoxal phosphate by Ehrlich ascites tumor cells. *Journal of Biological Chemistry, 236*:894.

Penrose, W. R., Nichoalds, G. E., Piperno, J. R., and Oxender, D. L. (1968): Purification and properties of a leucine-binding protein from *Escherichia coli. Journal of Biological Chemistry, 243*:5921.

Piperno, J. R., and Oxender, D. L. (1966): Amino acid-binding protein released from *Escherichia coli* by osmotic shock. *Journal of Biological Chemistry, 241*:5732.

Riggs, T. R., Coyne, B. A., and Christensen, H. N. (1953): Intensification of the cellular accumulation of amino acids by pyridoxal. *Biochimica et Biophysica Acta, 11*:303.

Riggs, T. R., and Walker, L. M. (1958): Diminished uptake of C^{14}-α-aminoisobutyric acid by tissues of vitamin B_6-deficient rats. *Journal of Biological Chemistry, 233*:132.

Riggs, T. R., Walker, L. M., and Christensen, H. N. (1958): Potassium migration and amino acid transport. *Journal of Biological Chemistry, 233*:1479.

Schultz, S. G., and Curran, P. F. (1970): Coupled transport of sodium and organic solutes. *Physiological Reviews, 50*:637.

Steiner, F. A. (1969): L-Glutamic acid, gamma-aminobutyric acid and pyridoxal 5′-phosphate at the level of the single unit in rat brain. In: *Vitamin B_6 in Metabolism of the Nervous System. Annals of the New York Academy of Sciences, 166*:199.

Stern, J. R., Eggleston, L. V., Hems, R., and Krebs, H. A. (1949): Accumulation of glutamic acid in isolated brain tissue. *Biochemical Journal, 44*:410.

Tager, H. S., and Christensen, H. N. (1971*a*): Hypoglycemic action of 2-aminonorbornane-2-carboxylic acid in the rat. *Biochemical and Biophysical Research Communications, 44*: 185.

Tager, H. S., and Christensen, H. N. (1971*b*): Transport of four isomers of 2-aminonorborane-2-carboxylic acid in selected mammalian systems and in *Escherichia coli. Journal of Biological Chemistry, 256* : 7572.

Tager, H. S., and Christensen, H. N. (1972): 2-Aminonorbornane-2-carboxylic acid; preparation, properties and identification of the four isomers. *Journal of the American Chemical Society,* (*in press*).

Tews, J. K., (1969): Pyridoxine deficiency and brain amino acids. *Annals of the New York Academy of Sciences, 166*:74.

Thomas, E. L., and Christensen, H. N. (1970): Indications of spatial relations among structures recognizing amino acids and Na^+ at a transport receptor site. *Biochemical and Biophysical Research Communications, 40*:277.

Thomas, E. L., and Christensen, H. N. (1971): Nature of the cosubstrate action of Na^+ and neutral amino acids in a transport system. *Journal of Biological Chemistry, 246* :1682.

Thomas, E. L., Shao, T. C., and Christensen, H. N. (1971): Structural selectivity in interaction of neutral amino acids and alkali-metal ions with a cationic amino acid transport system. *Journal of Biological Chemistry, 246*:1677.

Ueda, K., Akedo, H., and Suda, M. (1960): Intestinal absorption of amino acids. IV. Participation of pyridoxal phosphate in the active transfer of L-amino acids through the intestinal wall. *Journal of Biochemistry* (*Tokyo*), *48*:584.

Wheeler, K. P., and Christensen, H. N. (1967): Interdependent fluxes of amino acids and sodium ion in pigeon red blood cell. *Journal of Biological Chemistry, 242*:3782.

The Effect of Pyridoxine Deficiency on L-Aromatic Amino Acid Decarboxylase and Tyrosine Aminotransferase in Developing Brain

Samuel Eiduson, Arthur Yuwiler, and Elizabeth D. Eberle

Departments of Psychiatry and Biological Chemistry, Neuropsychiatric and Brain Research Institutes, UCLA Medical School, University of California, Los Angeles, California,
and
Neurobiochemistry Laboratory, Brentwood Hospital (Psychosocial Medicine), Veterans Administration, Los Angeles, California

For a number of years we have been concerned with various aspects of the developing brain (Eiduson, 1966; Eberle and Eiduson, 1968; Shih and Eiduson, 1969). Our interest has centered around the metabolism of the indolealkylamines and the catecholamines, especially with respect to the question of how the relevant substrates, enzymes, and products are regulated during early development. Although some workers have suggested end-product feedback inhibition of tyrosine hydroxylase in the case of norepinephrine biosynthesis (Costa and Neff, 1966; Spector, Gordon, Sjoerdsma, and Udenfriend, 1967), or end-product repression of tryptophan hydroxylase in the case of serotonin biosynthesis (Eiduson, 1966), little knowledge has been adduced concerning these regulatory mechanisms during the formative period of development. More recently, we have studied the developmental character of the isoenzymes of monoamine oxidase, and how their multiple forms may contribute to the regulatory process of the putative transmitters (Shih and Eiduson, 1969).

It has been suggested that pyridoxal-5′-phosphate (PLP) may be involved in the regulation of L-aromatic amino acid decarboxylase (Greengard and

Gordon, 1963). We wondered, therefore, about the effects cofactor availability or deficiency may have on the activity of this enzyme during maturation. Furthermore, we observed that when weaned rats were placed on a pyridoxine-deficient diet, the 5-hydroxytryptophan decarboxylase (5HTPD) activity of the liver was markedly decreased, whereas the decarboxylase activity of the brain was little affected (Wiss and Weber, 1964; Davis, 1963). Wiss and Weber (1964) demonstrated a reduction in 5HTPD activity in the brain of 15 to 20 day-old rats when the females nursing the young rats were placed on a pyridoxine-deficient diet from the day of parturition. These data suggested the importance of the stage of development of the infant rats considering the effect of pyridoxine deficiency on the 5HTPD activity in the brain. Therefore, we undertook to study the effect of pyridoxine deficiency on the aromatic L-amino acid decarboxylase activity in the developing brains and livers of young rats whose mothers were placed on a pyridoxine-deficient diet 2 weeks prior to mating and extending to 16 days post parturition. The experimental procedures were as described previously (Eberle and Eiduson, 1968). The

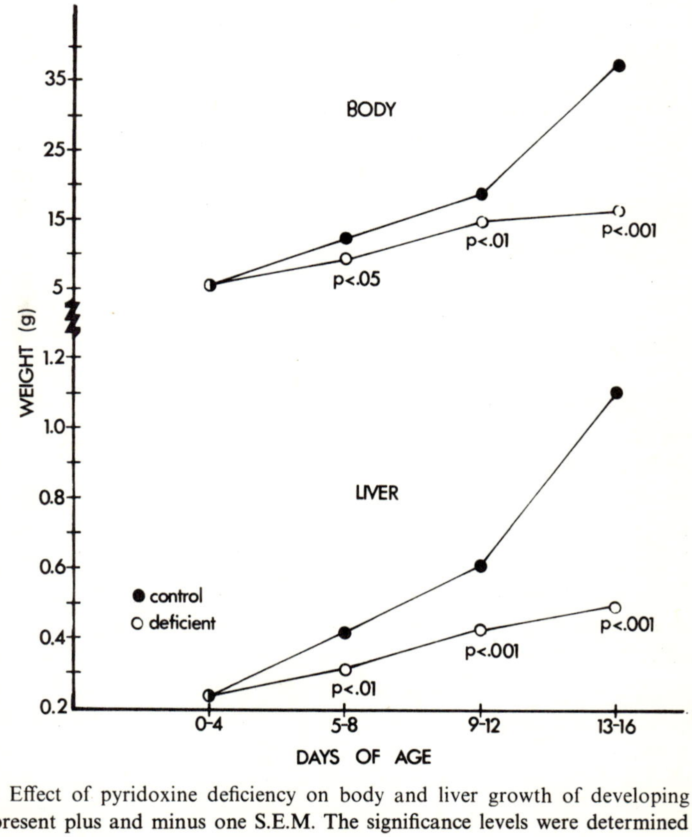

FIG. 1. Effect of pyridoxine deficiency on body and liver growth of developing rat. The bars represent plus and minus one S.E.M. The significance levels were determined by *t*-test.

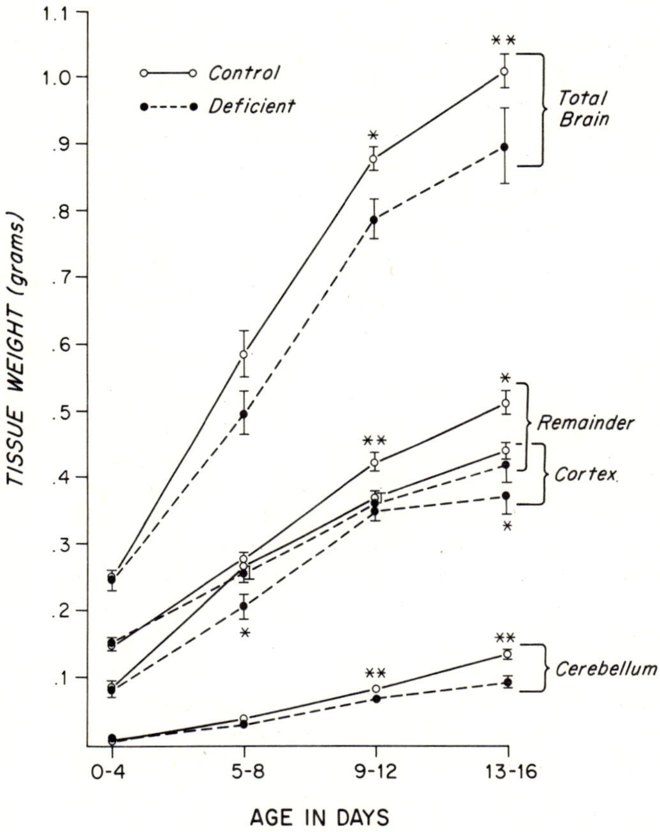

FIG. 2. Effect of pyridoxine deficiency on brain growth of developing rat. The symbols are the same as in Fig. 1.

activity of 5HTPD was measured using C^{14}-carboxy-labeled 5-hydroxytryptophan.

Comparison of body and liver weights of deficient and normal developing animals confirmed the previous observation that B_6 deficiency quickly retarded both body and liver growth (Fig. 1). The weights of all portions of the brain were lower in pyridoxine-deficient young rats (Fig. 2).

I. 5-HYDROXYTRYPTOPHAN DECARBOXYLASE DURING DEVELOPMENT

Liver: Maximal decarboxylase activity (enzyme measured with added PLP) of livers of young rats obtained from mothers on a normal diet showed a continuous rise from birth through the first 16 days. On the other hand, maximal decarboxylase activity in the deficient infant rats remained at its initially low level for the first 13 days and then dropped slightly. The activity of the holoenzyme as measured without added PLP also rose continuously from birth to 16 days of age in the control rats. By contrast, the activity of the holoenzyme in the deficient animals decreased over the 16-day period (Table 1).

Cerebellum: The maximal activity of the decarboxylase in the normal cerebellum increased very slowly during development, whereas that of the deficient animals decreased, reaching barely detectable levels at 16 days (Table 2).

Cortex: The control cortex had low but detectable 5HTPD levels at birth. These slowly increased until the middle of the second week at which time the increase in enzyme activity accelerated. By contrast, the maximal decarboxylase activity during deficiency soon fell lower than the control levels, and remained relatively constant. (Table 2).

Brain remainder (Table 2): In the deficient rats, the remainder of the brain had the highest levels of decarboxylase activity immediately following birth. The difference in enzyme activity between control and deficient rats at birth was not significant. However, a few days after birth, the maximal decarboxylase level of control rats continued to increase whereas that of the deficient rats decreased.

The data obtained from the above experiments provided evidence that the fetus was protected prenatally from the dietary B_6 deficiency imposed on the mother. It was equally clear that, while nursing from its B_6-deficient mother, the newborn rat was quickly and significantly affected, as judged from a marked slowing of growth rates in both body weight and liver. On the other hand,

TABLE 1. *5HTPD (μmoles) decarboxylated per gram wet weight per hour in liver*

Age (days)	Maximal enzyme (with PLP)		Holoenzyme (no PLP)	
	Control	Deficient	Control	Deficient
0–4	0.71 ± 0.04 (14)	0.58 ± 0.04[a] (20)	0.58 ± 0.04 (14)	0.41 ± 0.04[b] (20)
5–8	1.29 ± 0.04 (11)	0.63 ± 0.05[b] (16)	1.06 ± 0.04 (18)	0.29 ± 0.03[b] (18)
9–12	1.75 ± 0.16 (15)	0.71 ± 0.07[b] (21)	1.43 ± 0.11 (19)	0.30 ± 0.04[b] (23)
13–16	2.73 ± 0.13 (14)	0.41 ± 0.09[b] (9)	2.30 ± 0.13 (17)	0.15 ± 0.03[b] (9)

No. animals analyzed given in parentheses. Values ± S.E.M.

[a] $0.05 \leq P > 0.01$

[b] $P \leq 0.01$

TABLE 2. *5HTPD (μmoles) decarboxylated per gram wet weight per hour in cerebellum, cortex, and brain remainder*

Age (days)	Cerebellum				Cortex				Brain remainder			
	Maximal enzyme (with PLP)		Holoenzyme (no PLP)		Maximal enzyme (with PLP)		Holoenzyme (no PLP)		Maximal enzyme (with PLP)		Holoenzyme (no PLP)	
	Control	Deficient	Control	Deficient	Control	Deficient	Control	Deficient	Control	Deficient	Control	Deficient
0–4	.021 ± .009 (6)	.042 ± .012 (6)	.005 ± .000[a] (6)	.016 ± .007 (6)	.050 ± .004 (14)	.043 ± .003 (19)	.022 ± .003 (14)	.017 ± .002 (20)	.228 ± .010 (14)	.249 ± .005 (20)	.128 ± .009 (14)	.146 ± .008 (20)
5–8	.038 ± .006 (10)	.027 ± .005 (9)	.009 ± .003 (10)	.008 ± .002 (9)	.068 ± .005 (19)	.040 ± .003[b] (19)	.027 ± .003 (19)	.011 ± .002[b] (19)	.251 ± .008 (19)	.184 ± .007[b] (19)	.114 ± .011 (19)	.068 ± .009[b] (19)
9–12	.045 ± .004 (17)	.029 ± .003[c] (9)	.025 ± .003 (17)	.005 ± .000[b] (9)	.082 ± .004 (23)	.044 ± .003[b] (23)	.045 ± .004 (23)	.005 ± .000[b] (23)	.269 ± .010 (23)	.186 ± .007[b] (23)	.161 ± .010 (23)	.042 ± .005[b] (23)
13–16	.066 ± .003 (20)	.016 ± .003[b] (5)	.042 ± .003 (20)	.005 ± .000[b] (5)	.120 ± .005 (21)	.035 ± .007[c] (9)	.074 ± .005 (21)	.006 ± .001[b] (9)	.323 ± .010 (21)	.172 ± .012 (9)	.218 ± .013 (21)	.037 ± .007[b] (9)

No animals analyzed given in parentheses. Values ± S.E.M.
A value of 0.005 μmoles was arbitrarily assigned for statistical purposes whenever the assay gave negligible counts above background.
[b] $P \leq 0.01$
[c] $0.05 \leq P > 0.01$

although brain growth was eventually slowed, it was spared in its retardation for a longer period of time.

A comparison of the 5HTPD activity in the brain parts and livers of the young rats from mothers maintained on a pyridoxine-deficient diet showed that changes from normal in the liver enzyme were larger and were seen earlier than were changes in the different parts of the brain. The brain tissues also differed from each other: the control cerebellum exhibited the least enzyme activity, the brain 'remainder' had the highest activity, and the cortex exhibited intermediate activity.

The decarboxylase activity seemed to be affected comparably to tissue growth. The normal rate of increase in liver decarboxylase activity was much more affected, both in time and degree, than was the brain decarboxylase, again demonstrating the greater sparing effect in the brain than in the rest of the animal. The decarboxylase activity reported by Berman, Justice, and Hsia (1965) in kidneys of pyridoxine-deficient developing rats showed changes similar to the liver changes observed here.

The holoenzyme moiety of the 5HTPD was apparently most affected by the deficiency conditions employed in this study. The maximal enzymic activity (measured with added PLP) per gram of wet tissue merely remained unchanged in the deficient animal or dropped slightly with age instead of increasing as in the control animals. The apoenzyme moiety of the decarboxylase seemed generally to be maintained at control levels except for the brain remainder which showed higher apoenzyme levels in the deficient group after 9 days of age.

We wondered whether the deficient suckling rat could restore the lowered decarboxylase levels to normal if given sufficient cofactor, or whether it might remain deficient in enzyme protein and thus be unable to attain normal levels of enzyme activity. To test this question, female rats on a diet deficient in pyridoxine were mated, and the ensuing litters were allowed to nurse while their mothers continued on the deficient diet for 10 days, at which age the young rats were given daily doses of pyridoxine intraperitoneally (as was the mother). The activity of the aromatic amino acid decarboxylase was measured periodically for the ensuing 2 to $2\frac{1}{2}$ weeks.

II. WEIGHTS OF BODY, BRAIN AND LIVER

Body and weights (Fig. 3): Body weight increased rapidly during pyridoxine administration in all groups, although the deficient group continued to show a decrease in weight relative to the control group until 20 days of age (10 days of dosing). At this time, the growth of the deficient group apparently rose sharply while the rate of increase in the control group appeared to slow

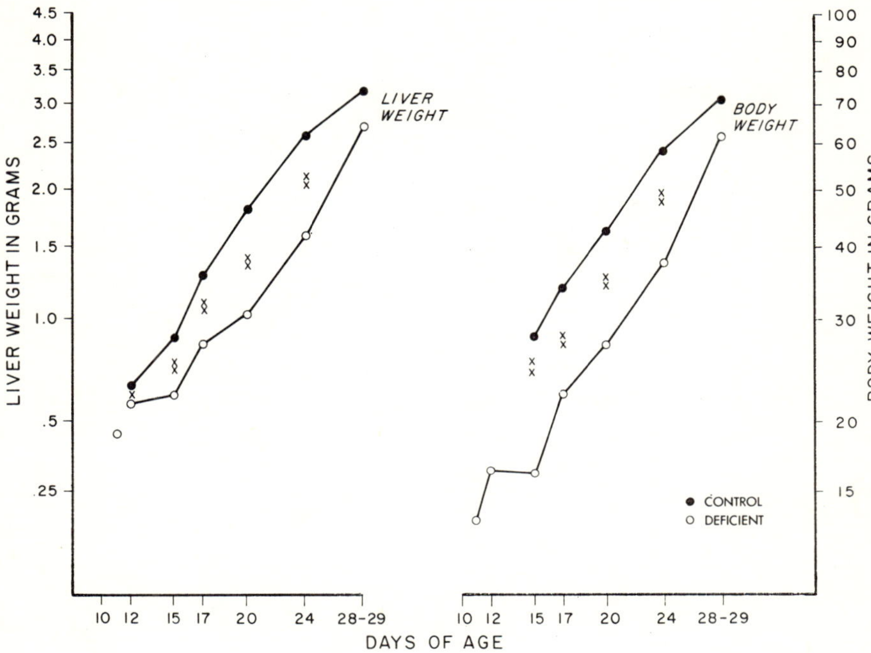

FIG. 3. Effect of pyridoxine administration to B_6-deficient rats on body and liver weights (see text for details). Upper curves represent data from control animals; lower curves show results from deficient animals. One "x" indicates $P < .05$; two "x" indicates $P < .01$.

down. The deficient animals had a lower liver weight than the control animals until after 24 days of age (2 weeks of pyridoxine administration). The control animals at 4 weeks showed a somewhat slower rate of increase in weight than did the deficient animals who received pyridoxine. Their growth continued at the previous rapid rate.

Brain (Fig. 4): The brains of the group receiving pyridoxine showed a statistically significant lower weight than the control group throughout the first 24 days (2 weeks of dosing), but no significant difference by 28 days of age. The weight of the cerebellum, however, remained significantly lower in the deficient group for the entire 4 weeks of study.

When the brain weight and liver weight were compared logarithmically to the body weight, there was an apparent linear relationship between the growth of the liver and the body regardless of the degree of pyridoxine deficiency, as was reported previously. Similarly, the brain weight to body weight relationship was not as apparent (Eberle and Eiduson, 1968).

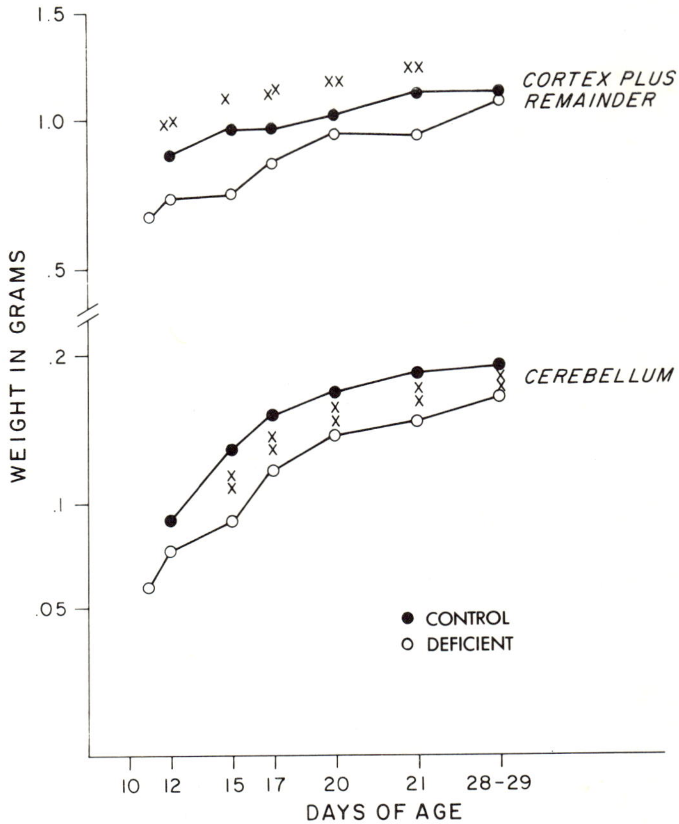

FIG. 4. Effect of pyridoxine administration on brain growth in B_6-deficient rats. The symbols are the same as in Fig. 3.

III. DECARBOXYLASE ACTIVITY

Liver (Table 3): When the decarboxylase activity per gram of wet weight of tissue was considered, the first change noted in the control group was the increase in percent saturation of the enzyme with cofactor after only 2 days of dosing. By 17 days of age (7 days of pyridoxine administration), there was no longer a statistically significant difference in the maximal amount of enzyme between the two groups, while the percent saturation with cofactor remained higher in the deficient group until after 24 days of age.

Brain (Tables 4 and 5): To clarify as well as reduce the data, the tables and graphs of the brain parts are presented for the cerebellum and for the cortex and remainder combined. The total brain decarboxylase concentration and content followed very closely the pattern of the combined cortex and remainder.

TABLE 3. *5HTPD (μmoles) decarboxylated per gram wet weight per hour in liver*

Age (days)	Maximal enzyme (with added PLP)		Holenzyme (no added PLP)		Apoenzyme (maximal minus hold)		% Saturation (holo/maximal × 100)	
	Control	Deficient	Control	Deficient	Control	Deficient	Control	Deficient
12	2.09 ± .09 (2)	.71 ± .11[a] (9)	1.05 ± .04 (2)	.52 ± .09[a] (9)	1.04 ± .05 (2)	.19 ± .05[a] (9)	50 ± 3 (2)	74 ± 7[a] (9)
15	2.17 ± .14 (3)	1.24 ± .20[a] (7)	1.51 ± .09 (3)	1.06 ± .17[b] (7)	.65 ± .05 (3)	.19 ± .05[a] (7)	70 ± 1 (3)	86 ± 2[a] (7)
17	2.18 ± .08 (4)	2.17 ± .12 (7)	1.89 ± .06 (4)	1.91 ± .11 (7)	.29 ± .03 (4)	.26 ± .04 (7)	87 ± 1 (4)	88 ± 1 (7)
20	2.60 ± .09 (5)	2.29 ± .10 (4)	2.04 ± .12 (5)	2.26 ± .09 (4)	.56 ± .11 (5)	.03 ± .07[a] (4)	79 ± 4 (5)	99 ± 3[a] (4)
24	4.17 ± .24 (6)	2.77 ± .25[a] (6)	3.30 ± .22 (6)	2.46 ± .25[b] (6)	.87 ± .06 (6)	.31 ± .08[a] (6)	79 ± 1 (6)	88 ± 3[b] (6)
28–29	3.53 ± .25 (4)	3.38 ± .15 (4)	2.99 ± .21 (4)	2.70 ± .12 (4)	.54 ± .08 (4)	.68 ± .08 (4)	85 ± 2 (4)	80 ± 2 (4)

No. animals analyzed given in parentheses. Values ± S.E.M.

[a] $P \leq 0.01$

[b] $0.05 \leq P$ 0.01 (between control and deficient)

TABLE 4. *5HTPD (μmoles) decarboxylated per gram wet weight per hour in cerebellum*

Age (days)	Maximal enzyme (with added PLP)		Holoenzyme (no added PLP)		Apoenzyme (maximal minus holo)		% Saturation (holo/maximal × 100)	
	Control	Deficient	Control	Deficient	Control	Deficient	Control	Deficient
12	.035 ± .005 (2)	.031 ± .004 (9)	.005 ± 0 (2)	.008 ± .002 (9)	.030 ± .005 (2)	.023 ± .005 (9)	15 ± 2 (2)	34 ± 10 (9)
15	.047 ± .002 (3)	.036 ± .004 (5)	.005 ± 0 (3)	.021 ± .005[a] (5)	.042 ± .002 (3)	.016 ± .003[b] (5)	11 ± 1 (3)	54 ± 12[a] (5)
17	.068 ± .003 (4)	.058 ± .002[a] (7)	.051 ± .003 (4)	.041 ± .004 (7)	.017 ± .002 (4)	.018 ± .004 (7)	75 ± 3 (4)	70 ± 8 (7)
20	.074 ± .005 (5)	.068 ± .012 (4)	.047 ± .001 (5)	.047 ± .009 (4)	.027 ± .004 (5)	.021 ± .009 (4)	64 ± 3 (5)	70 ± 9 (4)
24	.078 ± .007 (6)	.070 ± .005 (6)	.048 ± .006 (6)	.057 ± .005 (6)	.030 ± .003 (6)	.013 ± .006[a] (6)	62 ± 3 (6)	83 ± 7[a] (6)
28–29	.087 ± .004 (4)	.066 ± .005[a] (4)	.066 ± .004 (4)	.059 ± .002 (4)	.021 ± .002 (4)	.008 ± .005 (4)	76 ± 3 (4)	90 ± 8 (4)

No. animals analyzed given in parentheses. Values ± S.E.M.

[a] $0.05 \geq P \geq 0.01$ (between control and deficient)

[b] $P \leq 0.01$

TABLE 5. *5HTPD (µmoles) decarboxylated per gram wet weight per hour in cortex plus remainder*

Age (days)	Maximal enzyme (with added PLP)		Holoenzyme (no added PLP)		Apoenzyme (maximal minus holo)		% Saturation (holo/maximal × 100)	
	Control	Deficient	Control	Deficient	Control	Deficient	Control	Deficient
12	.169 ± .002 (2)	.147 ± .009[a] (9)	.053 ± .008 (2)	.073 ± .005 (9)	.116 ± .011 (2)	.074 ± .008 (9)	32 ± 5 (2)	50 ± 3 (9)
15	.194 ± .012 (3)	.168 ± .005 (7)	.080 ± .008 (3)	.128 ± .012[a] (7)	.113 ± .005 (3)	.039 ± .011 (7)	41 ± 2 (3)	77 ± 7[b] (7)
17	.241 ± .002 (4)	.228 ± .005[a] (6)	.191 ± .006 (4)	.174 ± .010 (6)	.050 ± .007 (4)	.054 ± .010 (6)	79 ± 3 (4)	76 ± 4 (6)
20	.267 ± .003 (5)	.222 ± .023 (4)	.171 ± .007 (5)	.171 ± .024 (4)	.096 ± .007 (5)	.051 ± .004[b] (4)	64 ± 3 (5)	76 ± 3[a] (4)
24	.292 ± .008 (6)	.272 ± .013 (6)	.173 ± .003 (6)	.215 ± .007[b] (6)	.119 ± .007 (6)	.057 ± .018 (6)	59 ± 2[a] (6)	80 ± 6[a] (6)
28–29	.290 ± .006 (4)	.263 ± .005[a] (4)	.226 ± .004 (4)	.201 ± .001[b] (4)	.064 ± .003 (4)	.062 ± .005 (4)	78 ± 1 (4)	77 ± 1 (4)

No. animals analyzed given in parentheses. Values ± S.E.M.

[a] $0.05 \leq P > 0.01$ (between control and deficient)

[b] $P \leq 0.01$

The cortex and remainder results were combined when they were found to have similar patterns of change. Although all parts of the brain except the cerebellum had less maximal enzyme per gram of wet weight (deficient group) during the first days of pyridoxine restoration, this difference largely disappeared by the fifth day of pyridoxine administration (15 days of age). Again, the enzyme became more saturated with cofactor, quickly reaching (and in some cases even overshooting) control levels of activity, after the first days of dosing. When expressed as level of maximal enzyme per entire brain (or per entire brain section), the differences between the deficient and control groups became more evident, since the weights of the deficient tissues were significantly less. The percent of the total enzyme in the holoenzyme form tended to remain higher in the brain of the deficient group than in the controls.

During the first week of pyridoxine administration, there was an apparent tendency in the deficient group for the maximal enzyme to reach normal concentration. Despite this, however, at the end of the 28-day testing period, there appeared to be a decreased amount of enzyme in each part of the brain, even while the percent of saturation was constant.

It seems probable that the aromatic amino acid decarboxylase was restored almost to control levels when pyridoxine was given in large daily doses to 10-day-old deficient nursing rats. The data also suggest that the growth of the body, liver, and brain, except for the cerebellum which was retarded during the deficiency, continued at a normal rate resulting in similar values in both groups by 4 weeks of age. The cerebellum has relatively low decarboxylase activity at all early ages studied when compared to the rest of the brain. It may be that this tissue remains relatively more affected by the deficiency than other areas of the brain where there is greater decarboxylase activity.

I would like to turn now, very briefly, to some recent work we have done concerning the effect of B_6 deficiency on the induction of the enzyme tyrosine aminotransferase. Because of the suggestion that cofactor removal of apoenzyme from enzyme-forming sites (Greengard and Gordon, 1963; Holten, Wicks, and Kenney, 1967) may be an obligatory step in enzyme induction, it might be expected that PLP-deficient animals would be unable to respond to inducers with normal elevation in tyrosine aminotransferase activity. To test this, 22-day-old male Long-Evans rats were treated for 4 days with Cosa-Terramycin in drinking water to reduce production of pyridoxal compounds by intestinal flora. One group was placed on a B_6-deficient diet and another on a normal diet. After 30 days, normal animals and B_6-deficient animals were either sacrificed or given an inducer (Solu-Cortef, tryptophan), and sacrificed 4 hr later. Some deficient animals were injected with pyridoxine (5 mg/kg) and sacrificed 2 hr later or switched to a normal diet for 4 days. Tyrosine and tryptophan aminotransferases were determined by the method of Lin, Pitt, Civen, and Knox (1958), with and without PLP (0.135 mM) added to the

TABLE 6. *Effect of hydrocortisone administration on normal and pyridoxine-deficient animals*

Variables		Control ($\bar{X} \pm$ S.E.)		Pyridoxine-deficient ($\bar{X} \pm$ S.E.)	
		Basal	Plus hydrocortisone	Basal	Plus hydrocortisone
body weight (gr)		(8) 182.3 ± 4.4	(8) 186.6 ± 5.5	(20) 88.5 ± 3.9[a]	(12) 87.5 ± 4.3
tyrosine					
aminotransferase	No PP	(8) 1.22 ± 0.15	(8) 6.96 ± 0.86[a]	(20) 0.21 ± 0.05[a]	(12) 0.53 ± 0.03[b]
(μmoles/min/g protein)	+ PP	(8) 10.07 ± 1.12	(8) 51.39 ± 4.66[a]	(20) 8.15 ± 0.953	(12) 31.98 ± 1.16[b]
tryptophan					
aminotransferase	No PP	(8) 0.222 ± 0.021	(8) 0.495 ± 0.033[a]	(7) 0.059 ± 0.013[a]	(5) 0.073 ± 0.008
(μmoles/min/g protein)	+ PP	(8) 0.209 ± 0.021	(8) 0.497 ± 0.037[a]	(5) 0.112 ± 0.016[a]	(3) 0.278 ± 0.096[b]
tryptophan					
peroxidase	No H	(8) 4.25 ± 0.46	(8) 31.40 ± 0.41[a]	(20) 5.39 ± 0.51	(12) 22.99 ± 3.46[b]
(μmoles/hr/g protein)	+ H	(8) 17.44 ± 3.25	(8) 78.79 ± 2.29[a]	(20) 15.83 ± 1.60	(12) 51.01 ± 7.23[b]

Numbers in parentheses are numbers of animals analyzed.

[a] Differs from normally fed controls by $P < 0.05$

[b] Differs from basal levels of deficient animals by $P < 0.05$

assay system. Tryptophan peroxidase was determined by the method of Knox and Auerbach (1955) with and without addition of hematin (0.15 μM) to the assay system. Proteins were determined by the method of Lowrey as described by Layne (1957).

As can be seen from Table 6, and as expected, pyridoxine deficiency resulted in a diminished cofactor saturation of the tyrosine aminotransferase. Total activity of the enzyme was not statistically altered however. On the other hand, tryptophan aminotransferase activity was lower in pyridoxine-deficient animals. This enzyme is normally fully saturated with cofactor, and little if any stimulation occurs in normal animals upon addition of B_6 to the system. As expected, activity of tryptophan peroxidase, a heme-requiring enzyme but which is also inducible by glucocorticoids, was unaffected by the deficient diet. Basal levels of the enzyme were unaffected with or without added hematin.

The effects of pyridoxine deficiency on tyrosine and tryptophan aminotransferase activities were freely reversible. Four days of normal diet following 30 days of a pyridoxine-deficient diet was sufficient to raise tyrosine aminotransferase activity to 1.64 ± 0.15 μmoles/min/g protein (mean) $\pm$ (S.E.) measured in the absence of PLP and to 14.05 ± 1.40 measured in the presence of PLP. These values do not differ from the corresponding values of 1.22 ± 0.15 and 10.07 ± 1.12 found in normally fed controls. Similarly, tryptophan aminotransferase levels after 4 days of normal diet were 0.219 ± 0.016 and 0.222 ± 0.010, respectively, as compared to control values of 0.222 ± 0.021 and 0.209 ± 0.021. Tryptophan pyrrolase activities were not affected.

IV. EFFECT OF INDUCERS

Normally fed animals responded to hydrocortisone injection with a five-fold rise in tyrosine aminotransferase activity and a two-fold rise in tryptophan aminotransferase activity. Pyridoxine-deficient animals responded to this inducer with a fourfold rise in total assayable tyrosine aminotransferase activity and a twofold rise in tryptophan aminotransferase activity. The fact that any holoenzyme increase was perceptible in the pyridoxine-deficient group suggests that the newly formed apoenzyme was able to compete successfully for the small amount of available PLP. Although the magnitude of the increases elicited by hydrocortisone was roughly comparable in the two groups, the final enzymic activities attained were significantly different. Although pyridoxine-deficient animals started at about the same basal levels of aminotransferase and peroxidase activity as controls, induced levels were somewhat lower, probably reflecting a general decrease in capacity for rapid protein synthesis in pyridoxine-deficient animals rather than a specific effect on these enzymes. This interpretation is supported by the difference in body weights between the two dietary groups and by the fact that tryptophan

TABLE 7. *Effect of pyridoxine, tryptophan, and allylisopropylacetamide (Ala) on enzyme levels in pyridoxine-deficient animals*

Enzyme		Pyridoxine[a]			Tryptophan[b]			Ala[c]		
		N	$\overline{X} \pm$ S.E.	%C[d]	N	$\overline{X} \pm$ S.E.	%C	N	$\overline{X} \pm$ S.E.	%C
tyrosine										
aminotransferase	No PP	(5)	2.46 ± 0.39	(1171)	(4)	0.52 ± 0.29[e]	(248)	(4)	0.35 ± 0.11[e]	(167)
(μmoles/min/g protein)	+ PP	(5)	10.71 ± 0.99[e]	(131)	(4)	22.67 ± 6.7[e]	(278)	(4)	22.84 ± 1.09[e]	(280)
tryptophan										
aminotransferase	No PP	(5)	0.153 ± 0.084	(259)	(—)			(—)		
(μmoles/min/g protein)	+ PP	(5)	0.154 ± 0.018	(137)	(—)			(—)		
tryptophan peroxidase	No H	(5)	8.62 ± 1.26[e]	(160)	(4)	18.83 ± 7.01[e]	(349)	(4)	8.16 ± 3.08[e]	(151)
(μmoles/hr/g protein)	+ H	(5)	21.56 ± 4.02	(136)	(4)	26.43 ± 12.39[e]	(167)	(4)	23.37 ± 8.37[e]	(147)

[a] Injected with pyridoxine (5 mg/kg i.p.) 2 hr before sacrifice
[b] Injected with L-tryptophan (1 gm/kg i.p.) 4 hr before sacrifice
[c] Injected with allylisopropylacetamide (400 mg/kg i.p.) 4 hr before sacrifice
[d] % of pyridoxine-deficient animals
[e] Differs from control by $P < 0.05$

peroxidase is a heme- rather than a pyridoxal-requiring enzyme. Induction appears insensitive to varying cofactor concentration.

Tryptophan and allylisopropylacetamide also elevated levels of tyrosine aminotransferase and tryptophan peroxidase in pyridoxine-deficient animals (Table 7). Greengard and Gordon (1963) have reported that a single injection of at least 1g/kg of pyridoxine was required to obtain induction of tyrosine aminotransferase. A doubling of activity was seen within 2 hr. In the present experiment, a sub-inducing dose (5 mg/kg) of pyridoxine was given 2 hr before sacrifice to one group of pyridoxine-deficient animals to determine the ease and rapidity with which the deficiency can be reversed. This injection of pyridoxine to deficient animals increased cofactor saturation of tyrosine aminotransferase above the level in normal controls without increasing the absolute levels of enzyme. Interestingly, the saturation of tryptophan peroxidase was also increased over that of controls although absolute levels were not changed. This increase in saturation is unexplained, but may be due to a sudden increase in heme formation.

The results presented here suggest that decreased cofactor availability is not accompanied by a decreased ability to form new cofactor—at least for tyrosine aminotransferase.

Despite severe pyridoxine depletion, as evidenced both by decreased growth and by markedly decreased cofactor saturation of these enzymes, both hydrocortisone administration and administration of tryptophan or allylisopropylacetamide increased tyrosine aminotransferase activity to nearly the same extent in pyridoxine-depleted animals as in normally fed controls. These results, of course, do not preclude the possibility that only some minimal amount of cofactor is required to permit induction.

Tryptophan aminotransferase activity even more clearly reflects the degree of avitaminosis in animals fed a pyridoxine-deficient diet. This enzyme is ordinarily fully saturated with cofactor in normally fed animals. In contrast, this enzyme in pyridoxine-deficient animals was 52% saturated under basal conditions and 25% saturated after induction. Nonetheless, its activity was also increased by hydrocortisone.

V. DISCUSSION

The apoenzyme repressor model advanced for cofactor induction would predict increased enzyme formation accompanying increased cofactor availability and, conversely, decreased formation of new enzyme accompanying decreased cofactor availability. The first prediction previously has been shown not to be the case for tryptophan peroxidase. Increases in heme saturation of this enzyme in adult animals following infusion of hematin (Jayo, Nelson, and Rose, 1964) or injection of the heme precursor γ-aminolevulinic acid

(Wetterberg, Yuwiler, and Geller, 1969) failed to increase total enzyme activity, although they markedly increased the degree of cofactor saturation. The results presented here suggest that the second predication is also not verified, at least for tyrosine aminotransferase. Despite severe pyridoxine depletion, as evidenced both by decreased growth and by markedly decreased cofactor saturation of the pyridoxal-dependent enzymes studied, administration of hydrocortisone, tryptophan, or allylisopropylactamide increased tyrosine aminotransferase activity to nearly the same extent in pyridoxine-depleted animals as in normally fed controls.

These results, of course, do not preclude the possibility that only some minimal amount of cofactor is required to permit induction. Despite initial treatment of the depleted group with oxytetracycline to minimize vitamin contributions from intestinal bacteria, and despite continued maintenance on a pyridoxine-free diet, it is clear that animals were not fully depleted of PLP. Whether this pyridoxine was furnished by regrown intestinal flora or represents residual vitamin stores sequestered over the 30 days of experimental diet is unknown. The data suggest, however, that control of induction by PLP is unlikely to be a major mechanism under physiological conditions in the mature rat.

Acknowledgments. This work was supported in part by NIH Grant HD 192702, and State of California DMH Grant 66–2–34.8. Computing assistance was obtained fron the UCLA Health Sciences Computing Facility sponsored by NIH Grant FR-3.

REFERENCES

Berman, J. L., Justice, P., and Hsia, D. Y. (1965): The metabolism of 5-hydroxytryptamine (serotonin) in the newborn. *Journal of Pediatrics, 67*:603.

Costa, E., and Neff, N. H. (1966): Isotopic and non-isotopic measurements of the rate of catecholamine biosynthesis. In: *Biochemistry and Pharmacology of the Basal Ganglia.* E. Costa, L. J. Côté, and M. D. Yahr, editors, p. 141. Raven Press, New York.

Davis, V. E. (1963): Effect of cortisone and thyroxine on aromatic amino acid decarboxylation. *Endocrinology, 72*:33.

Eberle, E. D., and Eiduson, S. (1968): Effect of pyridoxine deficiency on aromatic L-amino acid decarboxylase in the developing rat liver and brain. *Journal of Neurochemistry, 15*:1071.

Eiduson, S. (1966): 5-Hydroxytryptamine in the developing chick brain: Its normal and altered development and possible control by end-product repression. *Journal of Neurochemistry, 13*:923.

Greengard, O., and Gordon, M. (1963): The cofactor-mediated regulation of apoenzyme levels in animal tissues. *Journal of Biological Chemistry, 238*:3708.

Holten, D., Wicks, W. D., and Kenney, F. T. (1967): Studies on the role of vitamin B_6 derivatives in regulating tyrosine α-ketoglutarate transaminase activity *in vitro* and *in vivo. Journal of Biological Chemistry, 242*:1053.

Jayo, M. V., Nelson, J. F., and Rose, S. (1964): The effect of systemic administration of haematin on the tryptophan pyrrolase activity of rat liver. *Biochimical et Biophysica Acta, 92*:44.

Knox, W. E., and Auerbach, V. H. (1955): The hormonal control of tryptophan peroxidase in the rat. *Journal of Biological Chemistry, 214*:307.

Layne, E. (1957): *Methods in Enzymology,* vol. 3. S. P. Colowick and N. O. Kaplan, editors, p. 447. Academic Press, New York.

Lin, E. C. C., Pitt, B. M., Civen, M., and Knox, W. E. (1958): The assay of aromatic amino acid transaminations and keto acid oxidation by the enol borate-tautomerase method. *Journal of Biological Chemistry, 233*:668.

Shih, J. C., and Eiduson, S. (1969): Multiple forms of monoamine oxidase in the developing brain. *Nature, 224*:1309.

Spector, S., Gordon, R., Sjoerdsma, A., and Udenfriend, S. (1967): End product inhibition of tyrosine hydroxylase as a possible mechanism for regulation of norepinephrine synthesis. *Molecular Pharmacology, 3*:549.

Wetterberg, L., Yuwiler, A., and Geller, E. (1969): Tryptophan oxygenase changes following α-aminolevulinic acid administration in the rat. *Life Sciences, 8*:1047.

Wiss, O., and Weber, F. (1964): Biochemical pathology of vitamin B_6 deficiency. *Vitamins and Hormones, 22*:495.

Advances in Biochemical Psychopharmacology, Vol. 4.
Raven Press, New York © 1972.

Some Interrelationships of Pyridoxal Phosphate, Folic Acid, and Serine Metabolism in Brain

William F. Bridgers and L. D. McClain

The Neurosciences Program, University of Alabama, Birmingham, Alabama 35233

This conference and a similar one held two years ago at the New York Academy of Sciences (Kelsall, 1969) have summarized a great deal of information in support of the importance of pyridoxal phosphate in normal brain maturation, in transmitter metabolism, and in brain electrical activity. We should like to emphasize another aspect of this area which did not receive attention in the earlier conference, namely, the interrelationships among the metabolism of serine, pyridoxal phosphate, and folate in brain maturation and function.

We shall present our view of some of these interrelationships by briefly reviewing the evolution of some of our work over the past few years. A consideration of where we started and where we may be headed makes it apparent, as is often the case, that as analysis of the problem deepened, the concept of the principal objective changed. For example, we began these studies with the hope of gaining an understanding of the regulation of serine biosynthesis and utilization in developing brain; after our interest was aroused because of serine's position as an obligatory and central precursor of the complex lipids of myelin (Bridgers, 1970).

Thus, after we had worked out the pathway utilized by brain for the biosynthesis of serine (Bridgers, 1965), we turned to a consideration of its utilization, and carried out a long but fruitless search for the key to the regulation of its entry into complex lipids (Bridgers, 1970). However, we were more fortunate when we examined serine's utilization by brain for formation of some other products, specifically the one-carbon derivatives of folic acid (Bridgers, 1968). We found ourselves then pursuing the new goal of elucidating the complex interactions of folate, to which we may refer collectively as "regulation of one-carbon metabolism." After a still closer look at these interactions, we became aware that serine, pyridoxine, and folic acid might be involved in events at the synapse in some previously unexpected way. Therefore, after presentation of our findings, we shall conclude by summarizing the direct and indirect evidence in support of a possible role for these interactions in mood and affect, and in disease states such as schizophrenia.

I. SERINE BIOSYNTHESIS IN BRAIN

When we commenced these studies, two different pathways for serine biosynthesis had been reported in various tissues and organisms. The so-called phosphorylated route (Umbarger, Umbarger, and Siu, 1963; Pizer, 1963) branches from glycolysis at 3-phosphoglycerate. Oxidation of this compound, in a NAD^+-dependent reaction, leads to phosphohydroxypyruvate. This intermediate, via reaction with glutamate, yields phosphoserine. The pathway is completed by the dephosphorylation of phosphoserine by a specific phosphohydrolase. At that time it was believed by some that serine also might be synthesized via a comparable nonphosphorylated rout from D-glycerate (Sallach, 1956; Willis and Sallach, 1962).

We found that a cell-free extract of mouse brain incubated with NAD^+, fluoride, magnesium, ATP, and glutamic acid can convert radioactive glucose into serine (Bridgers, 1965). The omission of fluoride from the incubation resulted in the reduction of counts in serine and phosphoserine. We attributed this salutary effect of fluoride to its inhibition of enolase with a resultant conservation of phosphoglycerate. Further support for the role of phosphoglycerate as precursor of serine was the finding that the incorporation was strongly inhibited by iodoacetate, a potent inhibitor of glycolysis at the level of glyceraldehyde phosphate dehydrogenase, which is, of course, prior to the formation of phosphoglycerate.

Utilizing the technique of isotope competition, we demonstrated that the pathway in mouse brain from glucose to serine did indeed involve phosphoglycerate and phosphoserine. In these experiments, where the radioactivity in ^{14}C-glucose was converted to ^{14}C-phosphoserine and ^{14}C-serine, the presence of an excess of D-glycerate-3-phosphate reduced the incorporation into serine and phosphoserine by over 90%. On the other hand, a pool of DL-glycerate had no effect. Similarly, a pool of L-phosphoserine prevented the incorporation of radioactivity into serine, whereas a pool of L-serine did not decrease the labeling of phosphoserine, ruling out the possibility that serine is a precursor of phosphoserine. The results of all these experiments support the likelihood that the phosphorylated pathway is utilized by brain and that the non-phosphorylated pathway is not.

In the same study we showed that all the necessary enzymatic equipment for the phosphorylated pathway is present in mouse brain (Bridgers, 1965), and we studied in some detail the feedback inhibition of phosphoserine phosphatase by serine (Bridgers, 1967, 1969), for this appears to be a major control point.

We shall not give here the details of these reactions (which the interested reader will find in the publications cited above), but we should point out that the phosphohydroxypyruvate-glutamate transaminase, which results in production of phosphoserine, is, as expected, a pyridoxal phosphate-dependent step (Hirsch and Greenberg, 1967). Indeed, in a study of brain amino acid levels in pyridoxine-deficient mice, lowered serine levels were found; this was attributed to the reduced activity of this obligatory transamination step in serine biosynthesis (Tews and Lovell, 1967).

II. SERINE UTILIZATION IN BRAIN

Let us now turn from consideration of the brain's biosynthesis of serine to its utilization, and specifically to the pyridoxal phosphate-dependent interactions. Figure 1 shows the most important of these relationships. The β carbon of serine is the major source of the one-carbon derivatives of tetrahydrofolate (FH_4), via serine transhydroxymethylase, which is a pyridoxal phosphate-dependent reaction. Cystathionine synthetase, also a pyridoxal phosphate enzyme, competes for serine, especially in brain where cystathionine may play some important, but not yet understood, role. Homocysteine, the other reactant in cystathionine synthesis, plays another key role as a regenerator of free tetrahydrofolate through an irreversible reaction catalyzed by N^5-methyl tetrahydrofolate homocysteine methyl-transferase, yielding methionine, which in turn is the primary methyl donor (via S-adenosyl methionine) for a number of important processes in brain metabolism including, for example, catechol-O-methyl transferase.

Time will not permit discussion of the scanty information available on the regulation of these interactions, except to comment upon a few key points. The effector of the enzymes of the serine biosynthetic pathway, in terms of repression-derepression-like phenomena, is not serine, but appears to be methionine (Fallon, Davis, and Goyer, 1968). It is also undoubtedly of importance that serine transhydroxymethylase is strongly inhibited by N^5-methyl tetrahydrofolate (Schirch and Ropp, 1967). Furthermore, and more to the point, in considering brain biochemistry, it is intriguing that the violent spontaneous convulsions seen in rats after administration of homocysteine can be prevented by the prior injection of serine (Sprince *et al.*, 1969). It is possible that these and other observations to be discussed below are pointing toward brain-specific roles for these interactions, all of which are most probably involved in the regulation of folate metabolism.

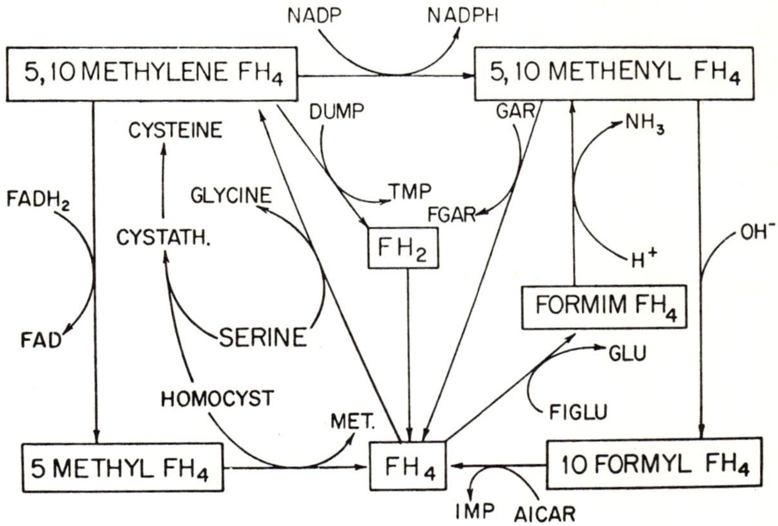

FIG. 1. The relationship of serine to folate coenzyme metabolism. The β-carbon of serine is a major source of N^{5-10}-methylene tetrahydrofolate (5,10-methylene FH_4) via serine transhydroxymethylase. Methylene FH_4 may be reduced to N^5-methyl tetrahydrofolate (5-methyl FH_4), oxidized to $N^{5,10}$-methenyl tetrahydrofolate (5,10-methenyl FH_4), or utilized for thymidylate (TMP) synthesis, forming dihydrofolate (FH_2). Tetrahydrofolate (FH_4) may be regenerated by reduction of FH_2, by methionine formation from 5-methyl FH_4, or from N^{10}-formyl tetrahydrofolate (10-formyl FH_4) as 5-aminoimidazole-4-carboxamide ribonucleotide (AICAR) is converted to 5-formamidoimidazole-4-carboxamide ribonucleotide (which ring then closes to yield inosinic acid). FH_4 is also consumed in histidine catabolism as formiminoglutamate (FIGLU) is converted to glutamate (GLU) and formiminotetrahydrofolate (FORMIN FH_4). Formin FH_4 may serve as a source of 5,10-methenyl FH_4, from which FH_4 may be regenerated when glycinamid ribonucleotide (GAR) is converted to N-formylglycinamide ribonucleotide (FGAR). In a closely related series of reactions, serine and homocysteine (HOMOCYST) yield cystathionine (CYSTATH), which is cleaved to form cysteine and α-ketobutyrate.

Let us consider, for example, brain serine transhydroxymethylase. We found that this activity declines during the early postnatal development stage in mouse brain (Bridgers, 1968) (Fig. 2). This decrease in the potential for synthesis of N^{5-10}-methylene tetrahydrofolate paralleled the daily decline in brain growth rate during the first week or so of postnatal life. Therefore, quite obviously, there is a rapidly changing folate one-carbon metabolism during this early maturation phase in brain. These findings are of special interest in the light of the recent discoveries (Arakawa, 1970) of three groups of infants with severe brain maldevelopment and mental retardation, each associated with a different deficiency of a folate interconverting enzyme. These probable inborn errors, which have been detected for N^5-methyl tetrahydrofolate homocysteine methyltransferase, for formiminotransferase, and for cyclohydrolase, afford dramatic evidence for the role of folate in cerebral maturation and function. Other subsequent studies in rats have shown that

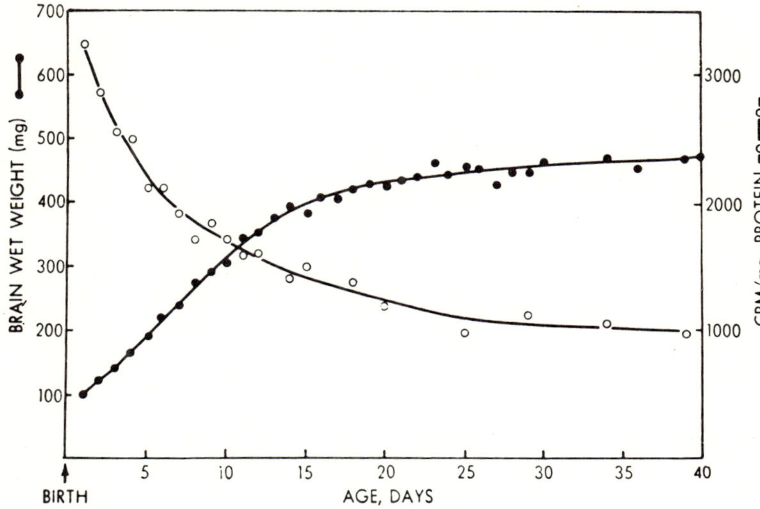

FIG. 2. Brain wet weight (•) in mg., and specific activity of serine transhydroxymethylase activity (o), measured as ^{14}C-glycine per mg protein, derived from incubation of brain extracts with ^{14}C-serine and tetrahydrofolate. For experimental details, see Bridgers (1968).

folic acid deficiency induced in the neonatal period causes a delay in maturation of the normal, spontaneous electrical activity of the brain, a change which may be irreversible (Arakawa *et al.*, 1969*a,b*).

III. BRAIN FOLATE DURING MATURATION

The possibility that there might be important changes in folate metabolism in developing brain led us next to a measure of folate coenzyme levels themselves (McClain and Bridgers, 1970). The concentrations of folate coenzymes in mammalian tissues are far too small to permit a direct chemical assessment. Fortunately, the growth responses of certain bacteria are sufficiently reliable to afford a microbiological method for assay. Figure 3 shows a measure of folate in extracts of mouse brain, based upon the growth responses of *Lactobacillus casei* and *Streptococcus faecalis.* These organisms are unable to utilize the naturally occurring poly-γ-glutamyl forms of folate, but *L. casei* can respond to folates containing three or fewer glutamyl residues, and *S. faecalis* can grow if supplied with folate containing two or one glutamyl moieties. The open triangle curve in Fig. 3 is the folate in brain extracts present as the mono-, di-, or triglutamyl derivatives. The solid triangle curve is obtained after prior treatment of the brain extracts with human plasma conjugase, a specific carboxypeptidase which sequentially hydrolyzes the glutamyl residues to the mono- or diglutamyl folates. Thus, the uppermost curve is a measure of

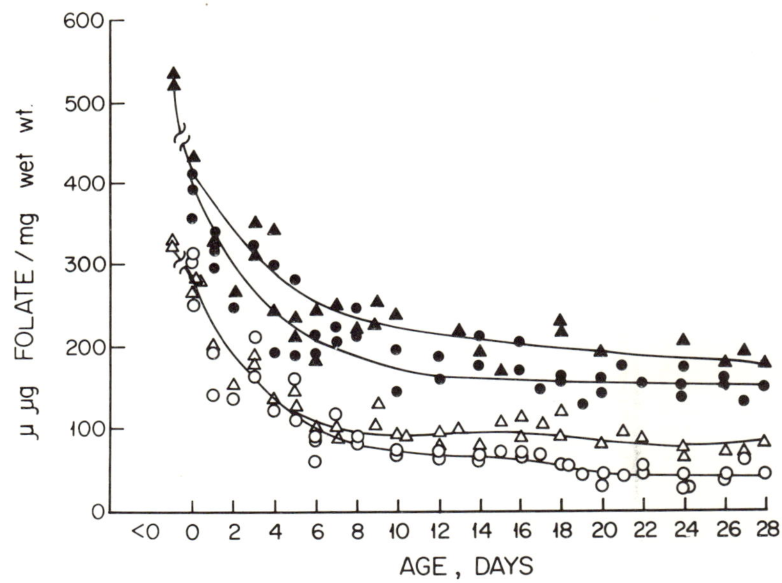

FIG. 3. Levels of folates in extracts of developing mouse brain as measured by the growth responses of *L. casei* without conjugase treatment (Δ) and after conjugase treatment (▲), and of *S. faecalis*, without conjugase treatment (o) and after conjugase treatment (●).

the total brain folate, and the difference between the two triangle curves would be a measure of the polyglutamyl forms. It is striking that the proportion of cerebral folate in the polyglutamyl forms is far less than has been found in other mammalian tissues or in yeast (Noronha and Silverman, 1962; Schertel, Boehne, and Libby, 1965; Kutzbach, Galloway, and Stokstad, 1967). It is also apparent that a smaller proportion of the total folate is in this polyglutamate form in the more immature brain, and that with maturation the folate concentrations declined as the proportion in polyglutamate forms increased. The solid and open circle curves represent brain folates, as measured by the growth response of *S. faecalis*, an organism which is unable to utilize N^5-methyl tetrahydrofolate. Thus, the closed circles represent the total folate, after conjugase treatment, except for the N^5-methyl forms, and the open circles represent the non-N^5-methyl folates present in the extract as mono- or diglutamyl derivatives. It should be noted that the total folates, most especially in the immature brains, were similar with both organisms, suggesting that much less of the folates in immature brain are the N^5-methyl derivative than is the case for the mature brain or for plasma, red cell, or liver folates (Buchanan, 1964; Silverman, Law, and Kaufman, 1961). We believe these results suggest that further exploration of the role of folate in brain maturation might be rewarding.

IV. COMPARTMENTATION OF BRAIN FOLATE

Let us now turn to the question of the intracellular localization of brain folates. For these experiments we have employed a standard fractionation procedure. Young adult mouse brain was homogenized in isotonic sucrose, and the nuclear fraction and cell debris were removed by a low-speed centrifugation. Sedimentation of a crude particulate fraction revealed that a surprisingly large proportion of the brain folate was found in this material. In fact, in several experiments the ratio of particulate to soluble folate has been found to be near unity. These findings in mouse brain stand in striking contrast to similar measurements in rat liver (Wang, Koch, and Stokstad, 1967) where only 6% of the total folate was found in the crude particulate fraction and over 90% was "cytoplasmic."

We next subfractionated the crude particulate material by discontinuous sucrose gradient centrifugation (Fiszer and De Robertis, 1967). This technique results in a separation of the particulate material into several undoubtedly heterogeneous fractions which, by the nomenclature of De Robertis, are labeled mitochondria, noncholinergic and cholinergic synaptosomes, and a heterogeneous fraction of disrupted synaptosomes and other membrane debris. Figure 4 presents our findings on folate levels within these compartments. The first bar in each set represents folate specific activity in grams $\times\ ^{-10}$ per milligram of protein. We noted above that the brain folate is split approximately evenly between soluble and particulate fractions, and here we see that the specific activities are also comparable. Furthermore, as shown in Fig. 4, the specific activity of folate in brain mitochondria was considerably lower than in the noncholinergic and cholinergic synaptosome fractions. It is of interest that the purified mitochondrial fraction contained essentially only N^5-methyl folates, much of which was in a polyglutamate form. Once again, the contrast with rat liver mitochondria is remarkable, for in that tissue not more than 30% of mitochondrial folate was in the methylated form, and approximately the same percentage was polyglutamates (Wang, Koch, and Stokstad, 1967).

Indeed, it is curious that in every fraction there appears to be a correlation between the amount of folate in the N^5-methyl form and the amount in the polyglutamate form. If further experiments bear out this possibility that folate polyglutamates are mostly the N^5-methyl derivative, and vice-versa, certain implications about the specificities of the folate conjugase and poly-γ-glutamyl synthetase systems will be apparent.

We have recently commenced a series of experiments to further localize the folate within the synaptosomes (Fig. 5). If one exposes the crude particulate material to water, a certain proportion of the synaptosomes are osmotically ruptured. In our experiments over 80% of the synaptosomal lactate dehydro-

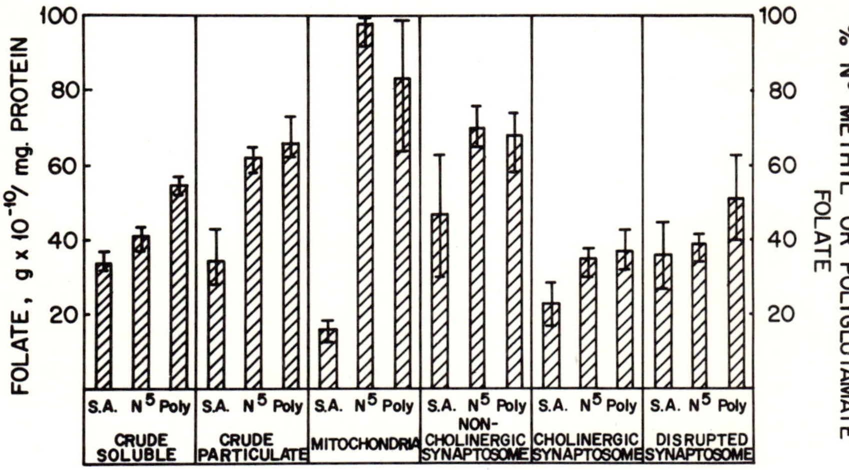

FIG. 4. Sucrose gradient centrifugation of a crude mitochondrial fraction from adult brain. The data represents the means and ranges from four separate preparations. Crude particulate = the material sedimenting from 1,100 × *g* to 27,000 × *g* in a 0.32 M sucrose homogenate of mouse brain. Crude soluble = supernatant material from above centifugation. First bar (S.A.) represents the specific activity of folate, as determined by *L.casei* assay, in g × 10^{-10} folate per mg protein. Second bar (N^5) represents % of total folate as the N^5-methyl derivative, as determined by differential growth of *L. casei* and *S. faecalis*. Third bar (Poly) represents % of total folate with more than two glutamyl residues, as determined by growth of *L. casei* with and without conjugase treatment of the extracts. See text for details of preparations and microbiological assays.

genese, a convenient, soluble cytoplasmic marker, is released after this treatment. However, only about 20% of the folate is released into a soluble form. Sedimentation of the residue from this treatment through a discontinuous sucrose gradient revealed a particulate fraction, migrating in the region where De Robertis finds synaptic ghosts or nerve ending membranes (Rodriguez, Alberici, and De Robertis, 1967; Fiszer and De Robertis, 1969), very greatly enriched with respect to folic acid, strongly supporting the likelihood that much of the synaptosome folate is bound to synaptosomal membrane.

Returning to the residue obtained after osmotic lysis of the crude particulate fractions, we find that an even more severe disruption with Triton-X-100 releases little more folate. De Robertis believes that this treatment results in the destruction of most of the synaptic membrane, leaving intact in the residue what he terms the "junctional complex," composed of a piece of presynaptic membrane attached to a piece of post-synaptic material and some of the subsynaptic web.

When we fractionate this residue on sucrose gradients (Fig. 5), we find an even higher specific activity material in the upper portions of the gradient. Indeed, in a further refinement of this gradient centrifugation technique, we are finding bands with higher specific activities than those shown in Fig. 5. We

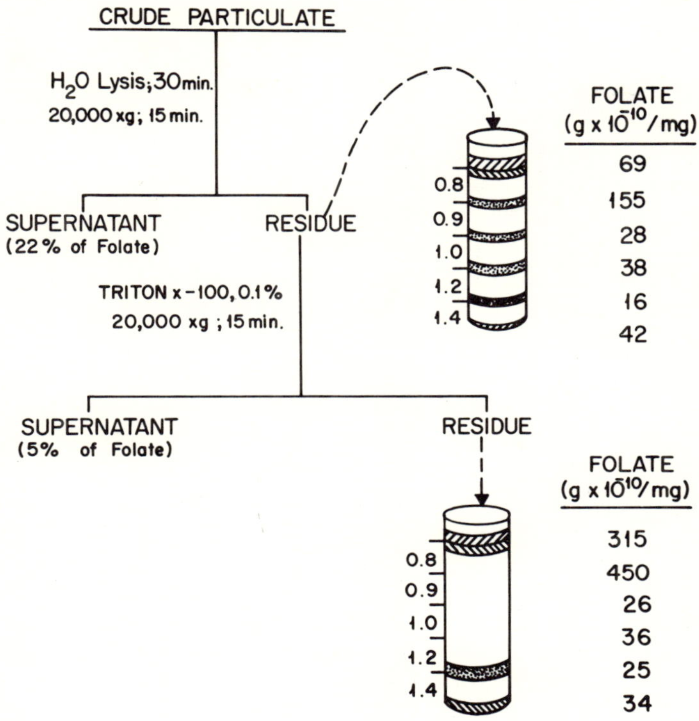

FIG. 5. Subfractionation of the crude post-nuclear particulate fraction of mouse brain by osmotic lysis and Triton-X-100 disruption, followed by discontinuous sucrose gradient centrifugation, as shown.

hasten to add that we do not yet have electron-microscopic pictures of this material; however, it is possible that this is the region where the junctional complexes migrate.

V. FOLIC ACID AND BRAIN FUNCTION

In closing, we mention some of the indirect evidence in support of an intimate relationship between folic acid and events at the synapse. This possibility is supported by the discovery that some patients on chronic anticonvulsant therapy are folate deficient, and exogenously administered folate may tend to remove the antiseizure effectiveness of these drugs in these patients (Chanarin, Laidlaw, Loughridge, and Mollin, 1960; Reynolds, Milner, Matthews, and Chanarin, 1966; Reynolds, 1967*b*, 1968). Additionally, there are some very interesting patients with an apparently congenital defect in absorption of folic acid. In these patients, who are mentally retarded and suffer from a seizure disorder, the frequency of convulsions is increased by folic acid administration and decreased when the patient is permitted to become folate deficient (Lanzkowsky, 1970).

It is also of interest that affective disturbances and organic psychoses are the typical psychiatric manifestations of folate deficiency (Reynolds, 1967*a*), and there are other reports of the immediate reversal by folic acid of a dementia associated with chronic anticonvulsant therapy (Anand, 1964).

In the light of the above discussion, and at the risk of being taken too seriously, we should like to note the possible relevance of some of these observations to the metabolic defect in schizophrenia. The most popular idea as to the basis of the defect in this condition is still the "transmethylation hypothesis" first put forward by Osmond, Harley-Mason, and Smythies (1952). This suggests that some abnormality in methylation (or demethylation) leads to the endogenous production of a methylated psychotogen chemically akin to mescaline or dimethyltryptamine. In support of this idea is the repeatedly confirmed finding that administration of methionine, a methyl donor, will cause an acute psychotic reaction in some schizophrenics (Pollin, Cardon, and Kety, 1961; Brune and Himwich, 1962; Sprince, Parker, Jameson, and Alexander, 1963; Park, Baldessarini, and Kety, 1965; Antun, Burnett, Cooper, Daly, Smythies, and Zealley, 1971). But serine is, like methionine, a good source of methyl groups, yet serine administration has not been reported to induce a psychotic reaction. Likewise, glycine, which was used as the control substance in some of these experiments (Pollin, Cardon, and Kety, 1961; Antun, Burnett, Cooper, Daly, Smythies, and Zealley, 1971) is nontoxic to schizophrenics, yet we now appreciate that it, too, is a good source of one carbon fragments (Yoshida, Kikuchi, Tada, Narisawa, and Arakawa, 1969). In fact, after extensive searches for the psychotogen there is still no experimental support for this hypothesis (Smythies, 1970).

Recognizing that the basis for the methionine effect is unknown, the possibility remains open that the effect may be some indirect sequelae of methionine administration. For example, we mentioned earlier the curious finding that methionine affects, by some unknown mechanism, the serine biosynthetic pathway, with high levels resulting in repression and low dietary levels resulting in derepression of the pathway. We also mentioned the observations that, for reasons that are still obscure, methionine administration results in a repartition of folate coenzyme forms (with a decline in the level of N^5-methyltetrahydrofolate, and therefore a concomitant rise in other folate forms). We suggested in a recent review (Bridgers, 1970) that these results are all compatible with a servo-type relationship between methionine, serine, and folate, mediated by some unknown effector, perhaps a specific folate form. We offer the suggestion that the methionine effect in schizophrenia deserves a reevaluation in the light of these relationships. When one couples these perturbations of folate with the indirect evidence for the role of folate in mood and affect, as discussed above, one might wonder whether a folate form is that elusive endogenous psychotogen.

REFERENCES

Anand, M. P. (1964): Iatrogenic megaloblastic anaemia with neurological complications. *Scottish Medical Journal, 9*:388.

Antun, F., Burnett, G., Cooper, A., Daly, R., Smythies, J., and Zealley, A. (1971): *In preparation.*

Arakawa, T. (1970): Congenital defects in folate utilization. *American Journal of Medicine, 48*:594.

Arakawa, T., Mizuno, T., Sakai, K., Chida, K., Watanabe, A., Ohara, K., and Coursin, D. B. (1969*a*): Electroencephalographic frequency patterns of rats treated with aminopterin in early infancy. *Tohoku Journal of Experimental Medicine, 97:*385.

Arakawa, T., Mizuno, T., Chida, M., and Narisawa, K. (1969*b*): Electroencephalogram and myelin lipids of six-week-old rats with aminopterin-treatment during suckling period. *Tohoku Journal of Experimental Medicine, 99*:59.

Bridgers, W. F. (1965): The biosynthesis of serine in mouse brain extracts. *Journal of Biological Chemistry 240*:4591.

Bridgers, W. F. (1967): Mouse brain phosphoserine phosphohydrolase and phosphotransferase. *Journal of Biological Chemistry, 242*:2080.

Bridgers, W. F. (1968): Serine transhydroxymethylase in developing mouse brain. *Journal of Neurochemistry, 15*:1325.

Bridgers, W. F. (1969): Purification of mouse brain phosphoserine phosphohydrolase and phosphotransferase. *Archives of Biochemistry and Biophysics, 133*:201.

Bridgers, W. F. (1970): The relationship of the metabolic regulation of serine to phospholipid and one-carbon metabolism. *International Journal of Biochemistry, 1*:495.

Brune, G. G., and Himwich, H. E. (1962): Effects of methionine loading on the behavior of schizophrenic patients. *Journal of Nervous and Mental Disease, 134*:447.

Buchanan, J. M. (1964): The function of vitamin B_{12} and folic acid coenzymes in mammalian cells. *Medicine, 43*:697.

Chanarin, I., Laidlaw, J., Loughridge, L. W., and Mollin, D. L. (1960): Megaloblastic anaemia due to phenobarbitone; the convulsant action of therapeutic doses of folic acid. *British Medical Journal, 1*:1099.

Fallon, H. J., Davis, J. L., and Goyer, R. A. (1968): Effect of protein intake on tissue amino acid levels and the enzymes of serine biosynthesis in the rat. *Journal of Nutrition, 96*:220.

Fiszer, S., and De Robertis, E. (1967): Action of Triton X-100 on ultrastructure and membrane-bound enzymes of isolated nerve endings from rat brain. *Brain Research, 5*:31.

Fiszer, S., and De Robertis, E. (1969): Subcellular distribution and chemical nature of the receptor for 5-hydroxytryptamine in the central nervous system. *Journal of Neurochemistry, 16*:1201.

Hirsch, H., and Greenberg, D. M. (1967): Studies on phosphoserine aminotransferase of sheep brain. *Journal of Biological Chemistry, 242*:2283.

Kelsall, M. A., Editor (1969): *Vitamin B_6 in Metabolism of the Nervous System. Annals of the New York Academy of Sciences, 166*:1–364.

Kutzbach, C., Galloway, E., and Stokstad, E. L. R. (1967): Influence of vitamin B_{12} and methionine on levels of folic acid compounds and folate enzymes in rat liver. *Proceedings of the Society for Experimental Biology and Medicine, 124*:801.

Lanzkowsky, P. (1970): Congenital malabsorption of folate. *American Journal of Medicine, 48*:580.

McClain, L. D., and Bridgers, W. F. (1970): Folate levels in developing mouse brain. *Journal of Neurochemistry, 17*:763.

Noronha, J. M., and Silverman, M. (1962): Distribution of folic acid derivatives in natural material. I. Chicken liver folates. *Journal of Biological Chemistry, 237*:3299.

Park, L. C., Baldessarini, R. J., and Kety, S. S. (1965): Methionine effects on chronic schizophrenics; patients treated with monoamine oxidase inhibitors. *Archives of General Psychiatry, 12*:346.

Pizer, L. I. (1963): The pathway and control of serine biosynthesis in *Escherichia coli. Journal of Biological Chemistry, 238*:3934.

Pollin, W., Cardon, P. V., Jr., and Kety, S. S. (1961): Effects of amino acid feedings in schizophrenic patients treated with iproniazid. *Science, 133*: 104.

Reynolds, E. H. (1967*a*): Vitamin B_{12} and folate deficiencies. *British Journal of Psychiatry, 113*: 681.

Reynolds, E. H. (1967*b*): Effects of folic acid on the mental state and fit-frequency of drug-treated epileptic patients. *Lancet, 1*: 1086.

Reynolds, E. H. (1968): Mental effects of anticonvulsants, and folic acid metabolism. *Brain, 91*: 197.

Reynolds, E. H., Milner, G., Matthews, D. M., and Chanarin, I. (1966): Anticonvulsant therapy, megaloblastic haemopoiesis and folic acid metabolism. *Quarterly Journal of Medicine, 35*: 521.

Rodriguez de Lores Arnaiz, G., Alberici, M., and De Robertis, E. (1967): Ultrastructural and enzymic studies of cholinergic and non-cholinergic synaptic membranes isolated from brain cortex. *Journal of Neurochemistry, 14*: 215.

Sallach, H. J. (1956): Formation of serine from hydroxypyruvate and L-alanine. *Journal of Biological Chemistry, 223*: 1101.

Schertel, M. E., Boehne, J. W., and Libby, D. A. (1965): Folic acid derivatives in yeast. *Journal of Biological Chemistry, 240*: 3154.

Schirch, L., and Ropp, M. (1967): Series transhydroxymethylase. Affinity of tetrahydrofolate compounds for the enzyme and enzyme-glycine complex. *Biochemistry, 6*: 253.

Silverman, M., Law, L. W., and Kaufman, B. (1961): The distribution of folic acid activities in lines of leukemic cells of the mouse. *Journal of Biological Chemistry, 236*: 2530.

Smythies, J. R. (1970): Schizophrenia and the mechanism of action of the hallucinogenic drugs. *Abstracts of World Medicine, 44*: 489.

Sprince, H., Parker, C. M., Jameson, D., and Alexander, F. (1963): Urinary indoles in schizophrenic and psychoneurotic patients after administration of tranylcypromine (parnate) and methionine or tryptophan. *Journal of Nervous and Mental Disease, 131*: 246.

Sprince, H., Parker, C. M., Josephs, J. A., Jr., and Magazino, J. (1969): Convulsant activity of homocysteine and other short-chain mercaptoacids: Protection therefrom. *Annals of the New York Academy of Sciences, 166*: 323.

Tews, J. K., and Lovell, R. A. (1967): The effect of a nutritional pyridoxine deficiency on free amino acids and related substances in mouse brain. *Journal of Neurochemistry, 14*: 1.

Umbarger, H. E., Umbarger, M. A., and Siu, P. M. L. (1963): Biosynthesis of serine in *Escherichia coli* and *Salmonella typhimurium. Journal of Bacteriology, 85*: 1431.

Wang, F. K., Koch, J., and Stokstad, E. L. R. (1967): Folate coenzyme pattern, folate-linked enzymes and methionine biosynthesis in rat liver mitochondria. *Biochemische Zeitschrift, 346*: 458.

Willis, J. E., and Sallach, H. J. (1962): Evidence for a mammalian D-glyceric dehydrogenase. *Journal of Biological Chemistry, 237*: 910.

Yoshida, T., Kikuchi, G., Tada, K., Narisawa, K., and Arakawa, T. (1969): Physiological significance of glycine cleavage system in human liver as revealed by the study of a case of hyperglycinemia. *Biochemical and Biophysical Research Communications, 35*: 577.

Advances in Biochemical Psychopharmacology, Vol. 4.
Raven Press, New York © 1972.

The Effect of Neuropharmacological Agents on Metabolism of Vitamin B_6 in Brain

E. E. McCoy, S. W. Henn, and C. E. Colombini

Departments of Pediatrics, University of Virginia, Charlottesville, Virginia, and University of Alberta, Edmonton, Alberta, Canada

The forms in which vitamin B_6 occur in mammalian brain (Bain and Williams, 1960) and the reactions of amino acid metabolism in which vitamin B_6 derivatives participate have been studied in detail (Snell and Ayling, 1968; Udenfriend, 1964; Braunstein, 1960).

Considerable work has been done on the characterization of enzymes involved in the phosphorylation (McCormick, Gregory, and Snell, 1961) and the oxidation (Pogell, 1958; Wada and Snell, 1961; Turner and Happold, 1961) of pyridoxine. Less information is available on the rates of metabolism of pyridoxine (PN)* in animals (Johansson, Lindstedt, and Register, 1966; Johansson, Lindstedt, and Tiselius, 1968), and only recently has a detailed study of vitamin B_6 metabolism in brain been published (Colombini and McCoy, 1970*a*) in which metabolic interconversion of all of the six forms was followed. Previous work has demonstrated that neuropharmacological agents that alter brain amine concentrations also alter the concentration of pyridoxal phosphate (PLP) (Smith, 1965; Bilodeau, 1965; Ebadi, 1970) or the activity of pyridoxal phosphokinase (Ebadi, Russell, and McCoy, 1968). It has also been shown that the concentration of PLP in brain affects the activity of pyridoxal phosphokinase (Ebadi, McCoy, and Kugel, 1970). Although the above data suggests that the concentration of amines and PLP exerts a control

* The following abbreviations are used throughout this article: PN, pyridoxine; PLP, pyridoxal phosphate; PL-kinase, pyridoxal kinase; PL, pyridoxal; PNP, pyridoxine phosphate; PMP, pyridoxamine phosphate; PM, pyridoxamine.

function on the metabolism of vitamin B_6 in brain, the detailed mechanisms of this regulatory system must be delineated.

The present paper reviews and extends work on the effects of neuropharmacological agents that alter brain amine concentration and affect the activity of pyridoxal phosphokinase. Data is also presented on the properties of purified rabbit brain pyridoxal phosphokinase, the normal metabolism of $PN^{14}C$, and the effect of marijuana, a hallucinogenic agent, on the metabolism of vitamin B_6 in mouse brain.

I. METHODS AND RESULTS

A. The Effects of reserpine, iproniazid, or DOPA on the activity of pyridoxal phosphokinase.

Details of drug administration, pyridoxal phosphokinase assay, and determination of amines were previously described (Ebadi, Russell, and McCoy, 1968). As shown in Fig. 1, the administration of reserpine resulted in approximately a 50% decrease in dopamine and norepinephrine and a 75% decrease in 5-HTP concentration. There was a concomitant fourfold increase in the activity of pyridoxal phosphokinase. When the monamine oxidase inhibitor iproniazid was administered, dopamine and serotonin concentrations in-

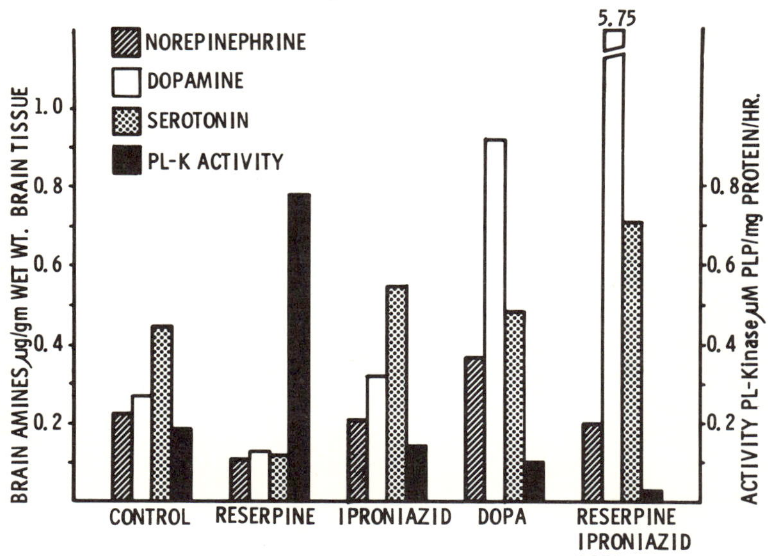

FIG. 1. Effect of neuropharmacological agents on the amine content and PL-kinase activity of rabbit brain. Reserpine (5 mg/kg) was given 16 hr prior to decapitation; DOPA (150 mg/kg) was given 2 hr prior to decapitation; reserpine (1 mg/kg) 16 hr; iproniazid (100 mg/kg) 8 hr; and DOPA (150 mg/kg) 2 hr prior to decapitation. Enzyme assays and amine determination as previously described (Ebadi, Russell, and McCoy, 1968).

creased approximately 20% while pyridoxal kinase (PL-kinase) activity decreased 20%. Very high brain amine concentrations achieved by the sequential administration of reserpine, iproniazid, and DOPA resulted in a 5.7-fold decrease in the activity of PL-kinase activity compared to controls. These studies indicated that *in vivo* an inverse relationship exists between the concentration of brain amines and the activity of PL-kinase.

B. Purification and properties of rabbit brain pyridoxal kinase (PL-kinase)

The *in vivo* studies with rabbits established that a rapid change in the activity of brain PL-kinase occurred with changes in the concentration of brain amines. The present *in vitro* studies were carried out to determine if various amines added to the incubation mixture would inhibit purified PL-kinase activity. PL-kinase was purified approximately 1,000-fold from rabbit brain according to the following procedure carried out sequentially: homogenization in 0.25 M sucrose, heating to 55°C in presence of 1mM pyridoxal (PL), and Sephadex G-200 chromatography. Protein was determined by the method of Lowry, Rosebrough, Farr, and Randall (1951), and the enzyme assayed by the method of Wada and Snell (1961). (A more detailed description of purification will be published later.) Assay of the eluents from the Sephadex G-200 column showed two major fractions for kinase activity (Fig. 2). Pre-

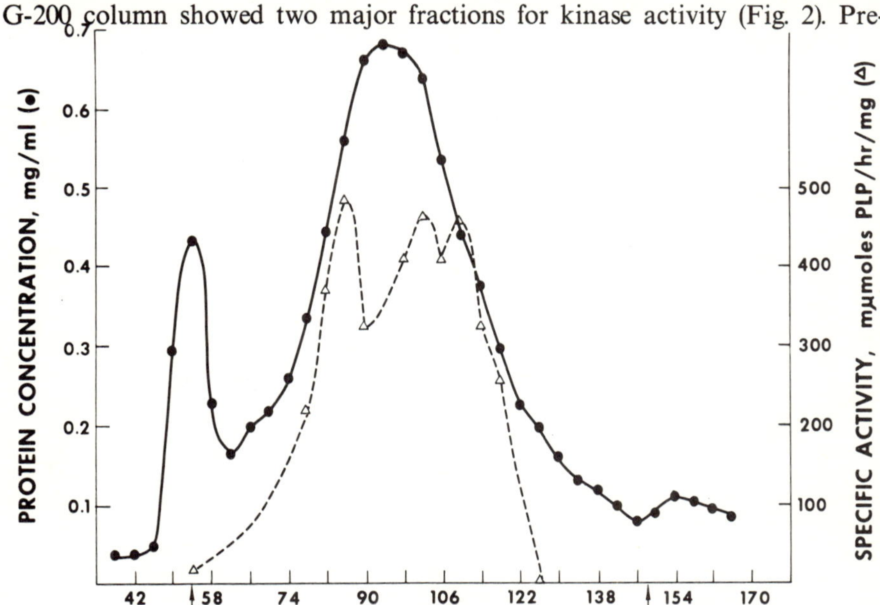

FIG. 2. Chromatography of purified rabbit brain PL-kinase on Sephadex G-200 column (2.5 × 37 cm). ●—● protein content of eluate, Δ----Δ PL-kinase activity. Methods for measurement of protein content and PL-kinase activity as in text.

liminary studies using a calibrated Sephadex G-200 column gave molecular weights of 94,000 and 44,000 for these peaks (Fig. 3). The purified enzyme of 94,000 mol. wt. was used to determine the effects of several amines on the activity of PL-kinase. The enzyme was assayed by the method of Wada and Snell (1961) in which the phenylhydrazone of PLP is determined in the presence of excess PL. As shown in Table 1, the addition of various amines at 1mM concentration resulted in an inhibition of kinase activity that varied from 35 to 87%. Of the compounds tested, the greatest inhibition was observed with

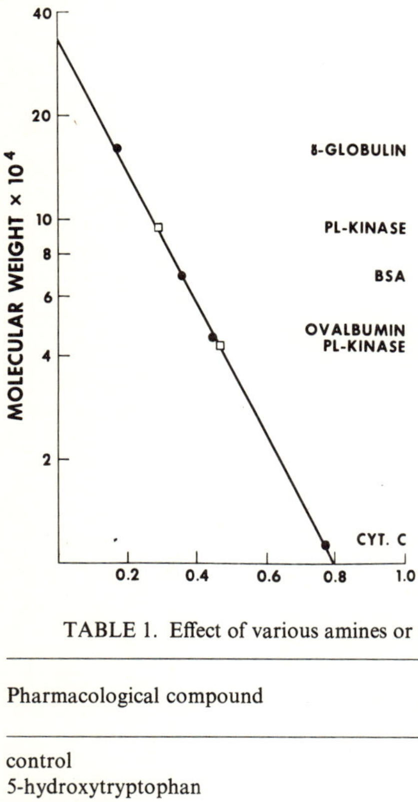

FIG. 3. Molecular weight determination of PL-kinase. Sephadex G-200 column was calibrated with proteins of known molecular weight as shown and logarithm of molecular weight *vs.* the molecular sieve coefficient σ plotted according to Whitaker (1963) and Andrews (1964). The molecular sieve coefficient of the two major Sephadex G-200 PL-kinase peaks was calculated and molecular weights determined from the plot.

TABLE 1. Effect of various amines or amine precursors on the activity of PL-kinase

Pharmacological compound	Specific activity (nmoles PLP/hr/mg)	Inhibition (%)
control	378	—
5-hydroxytryptophan	245	35
serotonin	95.6	75
DOPA	49.7	87
Dopamine	119	69
norepinephrine	119	69
norephedrine	84.1	78
melatonin	507	−34
pyridoxal semicarbazone	88.0	77

PL-kinase activity was determined by measuring the hydrazone of formed PLP in the absence or addition of an amine at 1 mM concentration. Results are expressed as percent inhibition compared to control PL-kinase activity.

DOPA (87%) and the least with 5-HTP. The secondary amine melatonin exerted a stimulatory action. A study of the mechanism of this inhibition is in progress.

C. The metabolism of pyridoxine ^{14}C in Brain

The studies previously noted demonstrated that changing amine concentrations affect the activity of PL-kinase and that *in vitro* addition of amines to purified PL-kinase inhibited the activity of this enzyme. In order to understand better the effect of varying amine concentration on B_6 metabolism in brain, we carried out an *in vivo* study of normal interconversions of vitamin B_6 in mouse brain. This study was undertaken by using PN-^{14}C of specific activity of 7 mC/mmole rather than a more highly labeled tritium labeled PN, because of the loss of tritium during the interconversion process. Swiss-Webster mice weighing approximately 29 g were injected intravenously with 5.3 μC PN^{14}C and sacrificed by immersion in liquid nitrogen at various intervals from 1 min to 7 days. Brains were removed while frozen and B_6 vitamers extracted and separated by methods previously described (Colombini and McCoy, 1970*b*). The brain retained 0.4%, liver 12%, and carcass 86% of the injected radioactive PN. The results in Fig. 4 are expressed as percent distribution of

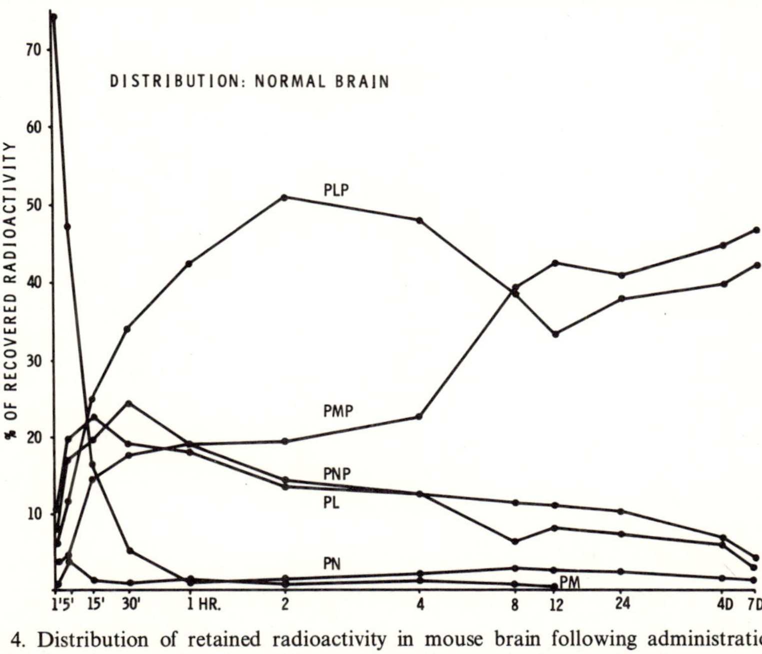

FIG. 4. Distribution of retained radioactivity in mouse brain following administration of PN-^{14}C. PN-^{14}C was administered i.v., and the mice sacrificed at various time intervals as shown. Brain isolation and separation of B_6 vitamers as described in text. Results are expressed as percent distribution among B_6 vitamers of retained radioactivity.

retained radioactivity of B_6 vitamers in brain at varying time intervals. Maximum retention was observed at 1 min after injection in liver (Fig. 5) and carcass (Fig. 6), but it did not occur until 5 min in brain. This delay in uptake of PN may be due to the blood-brain barrier. Injected PN was converted either to pyridoxal (PL) by pyridoxine oxidase or to pyridoxine phosphate (PNP) by pyridoxal phosphokinase. In subsequent steps, PL was phosphorylated to phosphate (PLP) and PNP oxidized to PLP. There then followed an interconversion of PLP to pyridoxamine phosphate (PMP) with an almost equal proportion of these compounds present after 24 hr. Five minutes after injection of $PN^{14}C$, when peak retention of radioactivity occurred, the distribution of labeled vitamers in brain was: PN 47%, PL 20%, PNP 17%, PLP 12%, and PMP 4%. After 15 min, the distribution was: PN 16%, PL 22%, PNP 20%, PLP 25%, PMP 14%, and PM 1.5%. After 2 hr the distribution was: PN 1.5%, PL 14%, PNP 14.5%, PLP 51%, and PMP 20%; after 24 hr the distribution was: PMP 41%, PLP 38%, PNP 10.5%, PL 7%, PN 2%, and PM < 1%. The results showed that injected PN-^{14}C was initially converted equally rapidly in brain to the active forms PL and PNP, which were then converted and retained primarily as PLP and PMP at the end of 24 hr. The initial distribution of retained radioactivity was different than that seen in liver or carcass.

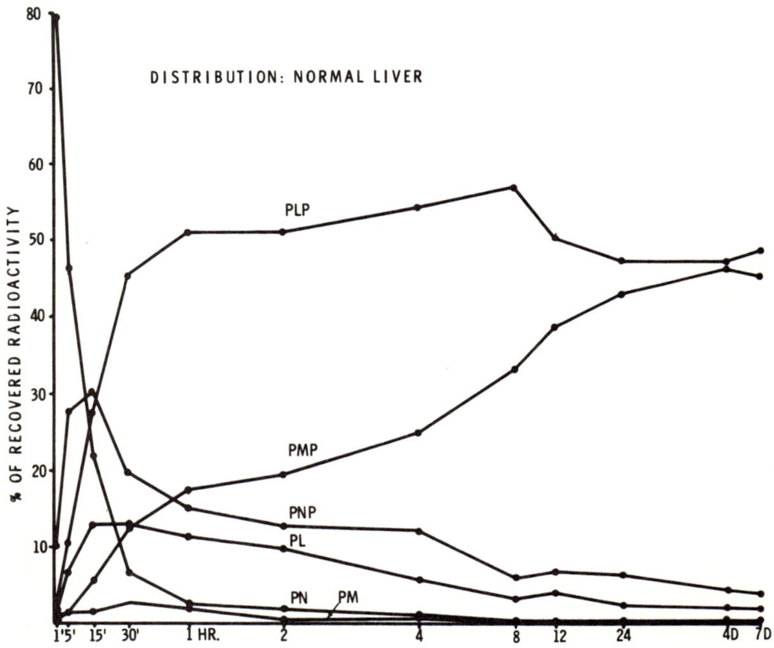

FIG. 5. Distribution of retained radioactivity in mouse liver following administration of PN-^{14}C to mice. Details similar to those described in Fig. 4.

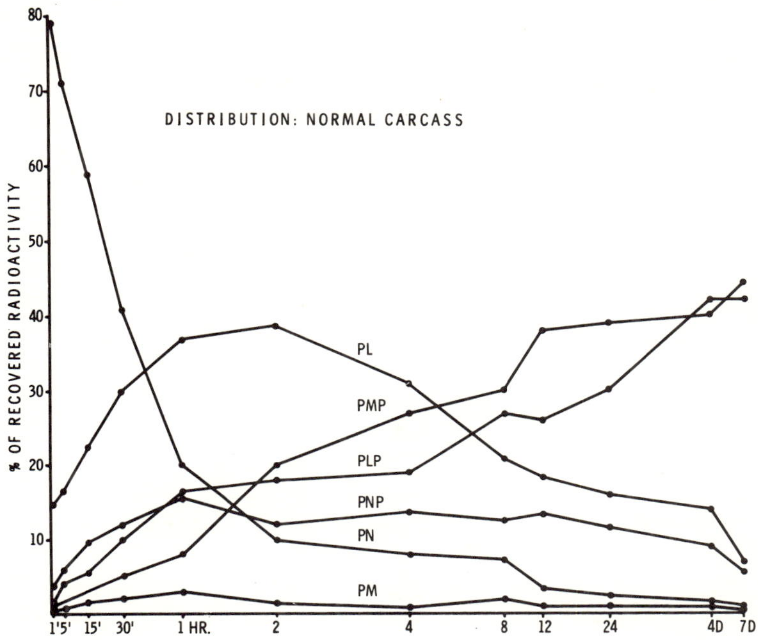

FIG. 6. Distribution of retained radioactivity in mouse carcass following administration of PN-^{14}C to mice. Details similar to those described in Fig. 4.

D. The effect of hallucinogens on vitamin B_6 metabolism in brain.

The previous section described the normal metabolism of PN-^{14}C in brain. The present section describes the effect of the commonly used hallucinogen, cannabis, on the metabolism of PN-^{14}C in mouse brain.

Tetrahydrocannabinol has been shown to increase the concentration of 5-HTP and decrease the concentration of norepinephrine in brain (Holtzman, Lovell, Jaffe, and Freedman, 1969). Studies were carried out on the metabolism of PN-^{14}C following a single injection of red oil extract of 500 μg/kg and after daily red oil injection of 300 μg/kg for 8 days. The isolation and separation of vitamers from brain has been described (Colombini and McCoy, 1970*b*). Following chronic administration of the red oil, there was increased retention of radioactive vitamers but a marked difference in distribution of retained radioactivity in brain. This was characterized by a threefold increase in the percentage of PNP from 13 to 39%, a reduction in PLP from 49 to 25%, a decrease in PL from 18 to 9%, and an increase in PMP from 14 to 21% (Table 2). The findings of an increased percent of PNP, decreased PL, and decreased PLP are consistent with an inhibition of pyridoxine oxidase following administration of the red oil of cannabis.

TABLE 2. Effect of marijuana administration on retention and distribution of B_6 vitamers in mouse brain

Tissue	Experiment	Total DPM	% of injected radioactivity	% PN	% PM	% PL	% PNP	% PMP	% PLP
brain	controls	9,571	0.16	3.17	2.17	17.98	13.03	14.17	49.57
	red oil (acute)	5,825	0.09	9.18	7.35	10.36	12.75	15.47	44.89
	red oil (8 days)	13,795	0.22	4.14	0.70	8.73	39.73	21.39	25.32
liver	controls	191,354	3.14	4.74	2.73	12.04	17.90	15.09	47.50
	red oil (acute)	243,433	3.99	1.53	0.53	13.99	7.64	14.69	61.58
	red oil (8 days)	231,687	3.80	12.00	0.26	12.91	31.23	11.72	31.81
carcass	controls	1,653,000	27.12	28.34	1.39	30.47	10.11	8.14	21.55
	red oil (acute)	1,197,100	24.58	11.45	3.15	43.68	13.33	11.49	16.90
	red oil (8 days)	2,058,200	33.77	24.55	1.08	57.59	4.93	6.38	5.47

Total radioactivity injected as PN-^{14}C = 6,093,000 DPM

Red oil containing 20% Δ^9-tetrahydrocannabinol injected in one dose of 500 μg/kg or daily for 8 days as described in the text. Animals sacrificed and B_6 vitamers separated as described. Results expressed as percent distribution of retained radioactivity.

II. DISCUSSION

This present work was undertaken to study some factors that affect the metabolism of vitamin B_6 in brain. PN, one of the principal sources of vitamin B_6 in man, is metabolized rapidly to its active form in two reactions involving phosphorylation and oxidation. Pyridoxal phosphokinase, the enzyme that phosphorylates all three forms of vitamin B_6, PN, PL, and PM, has been partially purified and characterized by McCormick, Gregory, and Snell (1961). The properties of pyridoxine oxidase, the enzyme that catalyzes the oxidation of either pyridoxine 5-phosphate or pyridoxamine 5-phosphate to pyridoxal 5-phosphate, have been studied by Wada and Snell (1961). When the available PLP is utilized in various biochemical reactions such as decarboxylation to form amines, its level should be restored either by oxidation of PMP and PNP or by phosphorylation of PL. The three amines dopamine, norepinephrine, and 5-HTP which have been implied to have neurotransmitter function (Rutledge and Weiner, 1967) require PLP for their formation. The effect of changes in amine concentration on the PL-kinase enzyme that phosphorylates PL has been studied by Ebadi, Russell, and McCoy (1968). It was demonstrated that several hours after the administration of drugs which increase or decrease brain amine concentration, there was a decrease or increase in the activity of PL-kinase. This would suggest that there exists a control mechanism that adapts to a decrease in amine concentration to increase PLP concentration for synthesis of new amines. The demonstration by Ebadi, McCoy, and Kugel (1970) that increased brain PLP concentration produced by administration of PN resulted in a decreased PL-kinase activity, whereas depletion of PLP by deoxypyridoxine resulted in an increase of PL-kinase activity, suggests another point of control on PL-kinase which involves the concentration of PLP. The relationships of each of these possible control mechanisms to the overall metabolism of PLP have yet to be established. The action of amines in the control of PL-kinase activity is not unique for it has recently been shown (Black and Axelrod, 1969) that the activity of hepatic tyrosine transaminase was inhibited by incubation with norepinephrine whereas depletion of norepinephrine was associated with increased activity of the enzyme.

PL-kinase was purified from rabbit brain and showed two peaks of molecular weight approximately 44,000 and 94,000 as determined by a calibrated Sephadex G-200 column. This is of interest in view of the demonstration of White and Dempsey (1970) that homogeneous PL-kinase from *Escherichia coli* has an estimated molecular weight of 40,000. The *in vitro* addition of an amine to the reaction mixture at 1 mM concentration resulted in inhibition of PL-kinase activity which was 75% for serotonin, and 69% for both dopamine and norepinephrine. Pyridoxal semicarbazone at 1mM concentration resulted in a 77% inhibition. This was less than the 93% inhibition observed by McCormick and Snell (1961) with pyridoxal semicarbazone at 0.5 mM concentration incubated with rat brain PL-kinase. In studies to be reported in detail elsewhere, incubation with 2×10^{-4}M DOPA resulted in 33% inhibition, whereas 1×10^{-3}M DOPA produced 86% inhibition of kinase activity. The development of a coupled assay to measure initial reaction will be necessary before it can be determined if the inhibition is competitive or non-competitive in nature and if it involves a Schiff-base formation. Incubation with melatonin resulted in an increase in PL-kinase activity. This is of particular interest in view of the finding (Antón-Tay, Sepulveda, and González, 1970) that intraperitoneal administration of melatonin increased *in vivo* brain PL-kinase activity. When these workers added melatonin to brain homogenate, no increase in activity was noted. We have found that the described effects of amines can be demonstrated only on a purified enzyme preparation. The mechanism of melatonin action on this enzyme is not known.

PN-^{14}C was used in the studies of metabolism of vitamin B_6 in mice because of its superiority over the tritiated compound in tracing the interconversions of this vitamin in various tissues. PN-^{14}C also gives better results than tritiated PN in the thin-layer electrophoretic separation of B_6 compounds in that there are much less streaking and better-defined spots to calculate percent recoveries. The maximum radioactivity retained was 86% in carcass and 12% in liver at 1 min, and 0.38% in brain at 5 min. The results in brain showed an increasing accumulation of PLP which peaked at 2 hr, and a slow rise in the amount and percentage of PMP which reached equilibrium with PLP 8 hr after the injection of PN-^{14}C. The examination of the results in brain showed that 5 min after the injection of labelled PN, 47% of the retained radioactivity was in the form of PN, 20% as PL, 17% as PNP, and 12% as PLP, whereas at 15 min there were equal amounts of PL, PNP, and PLP. These results would suggest that in brain PN has an equal opportunity either to be oxidized to PL or phosphorylated to PLP. Against the interpretation that this large percentage of PL is due to the action of pyridoxal phosphatase is the fact that with increasing time, there is a decreasing percentage of PL which would not be consistent if pyridoxal phosphatase were continuing to form the non-phosphorylated compound. The studies of Wada and Snell (1961) showed that pyridoxine phosphate oxidase had little activity on non-phosphorylated PN or PM. However, the present results would suggest that *in vivo* an active oxidase is present to convert PN to PL in brain. The data of Bain and Williams (1960) was also consistent with the present results. When PN was injected and phosphorylated and non-phosphorylated metabolites were isolated, it was found that there was a larger increase of PL than of PNP in the extracts of mouse brain 30 min later.

The administration of Δ^9-tetrahydrocannabinol resulted in a slight decrease in retention of radioactivity in brain and a change in distribution of B_6 vitamers, particularly noticed in an increased percentage of PNP. The acute administration of the crude red oil of cannabis caused a decreased retention in brain of injected PN, but after chronic administration the percent of radioactivity retained was greater than that of controls. Chronic administration was associated with a major shift in percent distribution of retained radioactivity—a threefold increase in PNP, a 50% decrease in PL, a 50% decrease in PLP, and an increase in PMP. These results are consistent with an inhibition of pyridoxine oxidase by the red oil of marijuana which resulted in a decreased conversion of PN to PL and an increase in both PNP and PMP which are dependent on the pyridoxine phosphate oxidase for their conversion to PLP.

The increase in the percent of PMP which occurred at the same time that PLP decreased cannot be explained by continued transamination of PLP to PMP but by inhibition of oxidation of free PMP to PLP. It remains to be determined if this altered distribution of B_6 metabolites would be sufficient to alter the rate of vitamin B_6-dependent reactions. Future work will involve the study of the effect of other neuropharmacological agents on the *in vivo* metabolism of vitamin B_6 in brain.

Acknowledgment. This work was supported by U.S. Public Health Service grants HD-03011 and ITD1-HD-00178.

REFERENCES

Andrews, P. (1964): Estimation of the molecular weights of proteins by Sephadex gel filtration. *Biochemical Journal, 91*:222.

Antón-Tay, P., Sepulveda, J., and González, S. (1970): Increase in brain pyridoxal phosphokinase activity following melatonin administration. *Life Sciences, 9*: part II, 1283.

Bain, J. A., and Williams, H. L. (1960): Concentrations of B_6 vitamers in tissues and tissue fluids. In: *Inhibitions of the Nervous System and Gamma-Aminobutyric Acid*, edited by E. Roberts, C. F. Baxter, A. Van Harreveld, C. Wiersma, W. Adey, and K. F. Killam, p. 275. Pergamon Press, New York.

Bilodeau, F. (1965): Effects of stimulants of the central nervous system on the pyridoxal phosphate content of the rat brain. *Journal of Neurochemistry, 12*:671.

Black, I. B., and Axelrod, J. (1969): Inhibition of tyrosine transaminase activity by norepinephrine. *Journal of Biological Chemistry, 244*:6124.

Braunstein, A. E. (1960): Pyridoxal phosphate. In: *The Enzymes*, Vol. 2, edited by P. D. Boyer, H. Lardy, and K. Myrback, p. 113. Academic Press, New York.

Colombini, C. E., and McCoy, E. E. (1970*a*): Vitamin B_6 metabolism. The utilization of pyridoxine-^{14}C by the normal mouse. *Biochemistry, 9*:533.

Colombini, C. E., and McCoy, E. E. (1970*b*): Rapid thin layer electrophoretic separation and estimation of all vitamin B_6 compounds and of some 5-hydroxyindoles. *Analytical Biochemistry, 34*:451.

Ebadi, M. S., Russell, R. L., and McCoy, E. E. (1968): The inverse relationship between the activity of pyridoxal kinase and the level of biogenic amines in rabbit brain. *Journal of Neurochemistry, 15*:659.

Ebadi, M. S., McCoy, E. E., and Kugel, R. B. (1970): Interrelationships between pyridoxal phosphate and pyridoxal kinase in rabbit brain. *Journal of Neurochemistry, 17*:941.

Ebadi, M. S. (1970): Increase in brain pyridoxal phosphate by chlorpromazine. *Pharmacology, 3*:97.

Holtzman, D., Lovell, R. A., Jaffe, J. H., and Freedman, D. X. (1969): 1-Δ^9-Tetrahydrocannabinol: Neurochemical and behavioral effects in the mouse. *Science, 163*:1464.

Johansson, S., Lindstedt, S., and Register, U. (1966): Metabolism of labeled pyridoxine in the rat. *American Journal of Physiology, 210*:1086.

Johansson, S., Lindstedt, S., and Tiselius, H. G. (1968): Metabolism of pyridoxine-$^3H_8$ in mice. *Biochemistry, 7*:2327.

Lowry, O. H., Rosebrough, N. J., Farr, A. L., and Randall, R. J. (1951): Protein measurement with the Folin phenol reagent. *Journal of Biological Chemistry, 193*:265.

McCormick, D. B., Gregory, M. E., and Snell, E. E. (1961): Pyridoxal phosphokinases: I. Assay, distribution, purification, and properties. *Journal of Biological Chemistry, 236*:2076.

McCormick, D. B., and Snell, E. E. (1961): Pyridoxal phosphokinases: II. Effects of inhibitors. *Journal of Biological Chemistry, 236*:2085.

Pogell, B. M. (1958): Enzymatic oxidation of pyridoxamine phosphate to pyridoxal phosphate in rabbit liver. *Journal of Biological Chemistry, 232*:761.

Rutledge, C. O., and Weiner, N. (1967): The effect of reserpine upon the synthesis of norepinephrine in the isolated rabbit heart. *Journal of Pharmacology and Experimental Therapeutics, 157*:290.

Smith, C. B. (1965): Effect of *d*-amphetamine upon brain amine content and locomotor activity in mice. *Journal of Pharmacology and Experimental Therapeutics, 147*:96.

Snell, E. E., and Ayling, J. E. (1968): Relationship of transaminase activity to structure of pyridoxal and its analogues. In: *Symposium on Pyridoxal Enzymes*, edited by K. Yamada, N. Katunuma, and H. Wada, p. 5. Maruzen Company, Tokyo.

Turner, J. M., and Happold, F. C. (1961): Pyridoxine phosphate oxidase and pyridoxal phosphatase activities in *Escherichia coli*. *Biochemical Journal, 78*:364.

Udenfriend, S. (1964): Amino acid decarboxylation steps in the biosynthesis of norepinephrine, serotonin, and histamine. *Vitamins and Hormones, 22*:445.

Wada, H., and Snell, E. E. (1961): The enzymatic oxidation of pyridoxine and pyridoxamine phosphates. *Journal of Biological Chemistry, 236*:2089.

Whitaker, J. R. (1963): Determination of molecular weight of proteins by gel filtration on Sephadex. *Analytical Chemistry, 35*:1950.

White, R. S., and Dempsey, W. B. (1970): Purification and properties of vitamin B_6 kinase from *E. coli. Biochemistry, 9*:4057.

Advances in Biochemical Psychopharmacology, Vol. 4.
Raven Press, New York © 1972.

Pyridoxine Dependency and Central Nervous System Excitability

Kurt Schlesinger and Lynda L. Uphouse

Department of Psychology and The Institute for Behavioral Genetics, University of Colorado, Boulder, Colorado

The role of pyridoxine (vitamin B_6) in the metabolism of amino acids, gamma-aminobutyric acid, biogenic amines, nucleic acids, proteins, and lipids is well known. Vitamin B_6 is involved in the normal functioning of the central nervous system, and its role in certain pathological processes of the brain is well documented. The purpose of this paper is to review some evidence which suggests that pyridoxine metabolism may be involved in determining susceptibility to sound-induced convulsions which are observed in certain inbred strains of mice. We shall also present data which suggests that gamma-aminobutyric acid (GABA), 5-hydroxytryptamine (5-HT, serotonin), and norepinephrine (NE) may be involved in determining susceptibility to sound-induced convulsions.

Susceptibility to sound-induced convulsions, or audiogenic seizures as they are usually called, is a phenomenon first observed in the rat by Donaldson (1924). Since then, the phenomenon has been described in animals of a variety of species including the mouse, *Peromyscus* (deermouse), rabbit, guinea pig, and man. In the main, research on audiogenic seizures has focused on biochemical, genetic, and pharmacological correlates of the phenomenon; very interestingly, little is known concerning the neuroanatomical and neurophysiological correlates underlying sound-induced convulsions. Nevertheless, an astonishing amount of research has accumulated on susceptibility to audiogenic seizures. This interest in audiogenic seizures is probably due to the following reasons. First, susceptibility to audiogenic seizures was originally believed to be a simple behavioral trait determined by a single genetic locus (Witt and Hall, 1949). Because of this, many investigators felt that it would be a relatively simple matter to determine and to understand some of the biological mechanisms underlying seizure susceptibility; perhaps a single enzymatic defect could be discovered which would adequately explain this inherited disorder. Further research, however, has revealed that the trait itself is quite complex, and that a single-gene hypothesis cannot account for audiogenic seizures (Fuller and Thompson, 1960; Ginsburg and Miller, 1963; Schlesinger, Elston, and Boggan, 1966). Although polygenic modes of inheritance of susceptibility to seizures are currently favored, a recent paper by Collins and Fuller (1969) indicates that single genes can be identified which contribute significantly to the expression of this trait. Also, mice of genotypes susceptible to audiogenic seizures tend to be more susceptible to other seizure-inducing agents than are mice resistant to sound-induced convulsions. On the assumption that susceptibility to seizures in general, and to audiogenic seizures in particular, is a measure of central nervous system excitability, it was thought that an understanding of some of the mechanisms which distinguish seizure-susceptible from seizure-resistant genotypes would go a long way toward an understanding of important central nervous system properties. In addition, there are certain similarities between susceptibility to audiogenic seizures and several central nervous system pathologies. We shall here list three such similarities. (1) Since certain mice susceptible to audiogenic seizures have been shown to be deficient in phenylalanine hydroxylase activity in liver, these mice might serve as a model for studying certain traits in human phenylketonuria; phenylketonurics, it should be pointed out, also show a very high incidence of epileptic seizures (Partington, 1961). (2) Susceptibility to audiogenic seizures has been shown to follow a circadian rhythm as, indeed, does the onset of epileptic seizures in human patients (Langdon-Down and Brain, 1929; Furuchi, 1969). This suggests that audiogenic seizures could be used as a model to study at least this aspect of epilepsy. (3) Certain patients, especially children, suffer from seizures which do not respond to the usual anticonvulsants. These patients have a greater evoked response to auditory stimulation than do control subjects. These patients also show an immediate and marked amelioration of their clinical symptoms following treatment with pyridoxine (Coursin, 1969). Susceptibility to audiogenic seizures also varies as a function of vitamin B_6 treatment; susceptibility to these seizures increases with pyridoxine deficiencies (Coleman and Schlesinger, 1965; Schlesinger and Schreiber, 1969) and decreases with the administration of supplemental vitamin B_6 (Schlesinger, 1968). Again, it seems possible that audiogenic seizures might be a useful model for studying the effects of vitamin B_6 on central nervous system functioning with particular reference to these types of cases.

These, then, are some of the reasons for the interest in sound-induced convulsions. The goal in all of this research is to understand the mechanisms which determine these inherited susceptibilities to audiogenic seizures. As has been pointed out previously (see, for example, Chance, 1963), we are a long way from this goal; we have, however, made some progress.

I. AUDIOGENIC SEIZURES

In our laboratory, audiogenic seizure tests are conducted in the following manner. Mice of different ages are taken from their home cages and are placed individually into a large chromatography jar (height 17 inches, diameter 11-$\frac{1}{2}$ inches) and given 30 sec to adapt. A 5-inch electric bell, mounted over the jar, is then sounded for 90 sec. The bell generates approximately 117 $\pm$ 3 db's of noise at the level of the mouse. The behavior of the animal in response to this intense acoustic stimulation is then observed and recorded. Many variations of this testing procedure have been employed in other laboratories; the major differences in these procedures are the intensity of the sound stimulus and the number of times each animal is used. In our laboratory, each animal is tested only once.

Sound-induced convulsions are a series of psychomotor reactions to an intense acoustic stimulus. The series of reactions observed during a "typical seizure episode" are the following. (1) A short latency period during which the animal may huddle at the edges of the jar, or wash and groom excessively. This latency period is of variable duration, with an average latency of approximately 3 sec and a range of from 1 to 20 sec. (2) A wild running phase which is characterized by frenzied running around the boundary of the container. This wild running fit appears to be rather characteristic of sound-induced convulsions and most clearly distinguishes audiogenic seizures from seizures induced by other means. (3) A clonic-flexor seizure during which the animal falls on its side while drawing its hind-limbs towards its chin. (4) A tonic-extensor seizure during which all four legs are extended caudally. (5) Death due to respiratory failure. Most animals can be revived by artificial respiration.

Many variations of this pattern have been described. As we have already noted, the latency period is of variable duration. The wild running phase may or may not be accompanied by a change of gait which is best described as a series of stiff-legged bounds. The episode may terminate at the wild running phase without progressing to a clonic-flexor seizure. The clonic seizure may terminate the episode without becoming a tonic seizure, and this seizure may or may not be fatal. Nevertheless, this seizure pattern is more or less characteristic and the order of events is nearly always the same. This makes the scoring of the response a relatively simple matter, and a number of different scoring procedures have been used. For example, each phase of the response may be graded one point, the latency period may be timed, the severity of the seizure may be judged and rated, the incidence of fatal seizures may be recorded, or other criteria may be used. Variations in testing procedure and scoring of the response in different laboratories have led to some confusion in the literature on audiogenic seizures. Since audiogenic seizures have now become a rather standard bioassay procedure, it is becoming increasingly important that criteria may be used. Variations in testing procedure and scoring of the response in different laboratories have led to some confusion in the literature on audiogenic seizures. Since audiogenic seizures have now become a rather standard bioassay procedure, it is becoming increasingly more important that testing and scoring procedures be standardized. Failing this, it is very important that the testing procedure be described in some detail in all experimental reports and that the exact criterion used to judge if a seizure has occurred be explained.

For this reason, we have usually reported the incidence of each phase of the response. For summary purposes we have used a "seizure severity score" in which each animal is given a score based on its response: no response = 0; wild running only = 1; wild running plus clonic seizure = 2; wild running plus clonic and tonic seizures = 3; wild running plus clonic, tonic, and lethal seizures = 4.

II. GENETIC FACTORS IN SUSCEPTIBILITY TO AUDIOGENIC SEIZURES

The importance of genetic factors in the determination of susceptibility to audiogenic seizures has been demonstrated in many laboratories and in animals of several species. Most of these data have been gathered on *Mus musculus*, although some very interesting work has also been done on the rabbit. Two general procedures are characteristic of this research, namely selective breeding experiments and strain comparison studies.

Frings and Frings (1953) have successfully selected for high and low susceptibility to audiogenic seizures in the mouse. The results of numerous strain-comparison studies on susceptibility to audiogenic seizures in mice indicate that seizure risk and seizure severity depend on the genotype of the animal being tested. Fuller and Thompson (1960) and Fuller and Wimer (1966) have reviewed many of these experiments. Nellhaus (1963) has reported the results of an experiment in which he developed lines of rabbits which were either susceptible or not susceptible to audiogenic seizures.

The animals most often used in research on audiogenic seizures are inbred DBA/2J and C57BL/6J mice; DBA/2J mice are susceptible to audiogenic seizures, whereas C57BL/6J mice are resistant to these seizures. DBA/2J mice are an example of a dilute strain, i.e., mice of dilute coat color which are homozygous for the autosomal recessive allele *d* in linkage group II. It is an interesting fact that mice of all dilute strains ever tested, i.e., DBA/2J, DBA/1J, BDP/J, P/J, I/FnLn, and DBA/Crg1, are susceptible to audiogenic seizures. It is also of interest that dilute lethal mice (d^l/d^l) suffer from spontaneous seizures. Finally, the rabbits susceptible to audiogenic seizures in Nellhaus' experiment were blue-eyed. In other words, dilute animals seem to be peculiarly susceptible to audiogenic seizures, although this is not meant to imply that only dilute animals are seizure susceptible. Indeed, many mice of non-dilute genotypes are susceptible to audiogenic seizures (Fuller and Sjursen, 1967). The evidence is also compelling that the dilute locus itself is unrelated to the expression of this trait (Schlesinger *et al.*, 1966). Nevertheless, it remains an interesting observation that mice of dilute genotypes are uniformly susceptible to audiogenic seizures, and this raises the possibility that some gene on linkage group II, linked to the dilute gene, mediates some component of the convulsive response.

The results of an experiment in which mice of various genotypes were tested for susceptibility to audiogenic seizures are summarized in Fig. 1. In this experiment, DBA/2J, C57BL/6J, and F_1 hybrid mice were used as experimental subjects; the animals were 14, 21, 28, 35, or 42 days of age at the time of testing. As can be seen, DBA/2J mice are susceptible to audiogenic seizures, whereas C57BL/6J mice are resistant to these seizures; F_1 hybrid mice are intermediate in seizure susceptibility although closer to the non-susceptible parent strains. An additional fact is clear from these data; the age of the animal at the time of testing is a very important variable in determining seizure susceptibility. Of the 21-day-old DBA/2J animals, 87% have clonic-tonic convulsions; this figure is fairly constant from experiment to experiment, ranging from 80 to 100%. On the other hand, only 13% of the 14- and 28-day-old DBA/2J mice tested have clonic-tonic seizures; this figure is also reliable with reports of clonic-tonic seizure incidence in animals of these ages ranging between 10 and 30%. Such age-dependent seizure susceptibility has been reported by other investigators. Vicari (1951) and Swinyard, Castellion, Fink, and Goodman (1963) have reported differences in seizure incidence and seizure severity in DBA/2J and O'Grady susceptible mice, respectively, as a function of age. The exact age of maximal seizure risk and maximal seizure severity depends on the genotype of the animals tested. However, even within inbred strains, different developmental patterns have been reported; this suggests that unspecified environmental variables such as diet, conditions of testing and housing, temperature, and/or time of day at which the seizure tests are conducted are important in determining seizure susceptibility. The developmental pattern of seizure susceptibility observed in F_1 hybrid mice is also illustrated in Fig. 1 and is similar to that observed in the DBA/2J mice although the hybrids are less seizure susceptible.

Table 1 summarizes the results of a similar experiment in which DBA/2J, C57BL/6J, F_1, F_2, DBA/2J $\times$ F_1, and C57BL/6J $\times$ F_1 mice were tested for susceptibility to audiogenic seizures. Data on 21-day-old animals are given, although similar experiments have been performed in 28-day-old mice (Schlesinger *et al.*, 1966). Numbers of animals showing wild running, clonic, or tonic seizures are given; the numbers in parentheses are expected values for a one-locus, two-allele hypothesis. These observations do not fit a single-gene

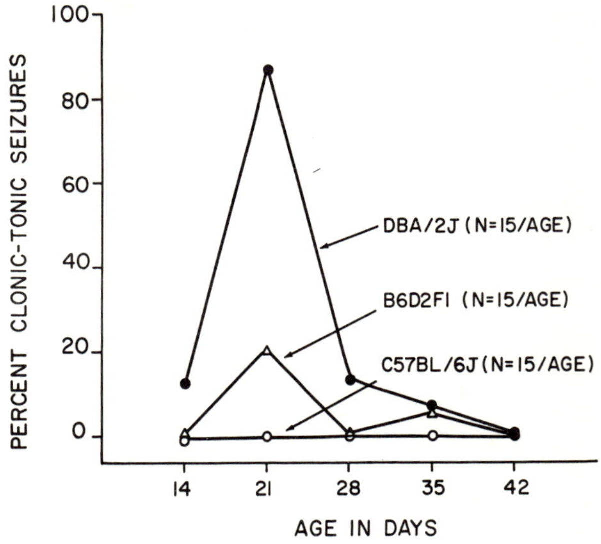

FIG. 1. Effects of genotype and age on susceptibility to audiogenic seizures (N = 15/age).

TABLE 1. *Number of 21-day-old mice responding by wild running, clonic, and tonic seizures to high intensity acoustic stimulation*

Genotype	Number tested	Numbers responding[a] Wild running	Clonic	Tonic
DBA/2J	37	36 (37.00)	33 (37.00)	34 (35.10)
DBA/2J × F_1	60	54 (50.52)	47 (33.84)	39 (30.87)
F_1	15	12 (10.26)	3 (1.92)	3 (1.25)
F_2	60	40 (35.52)	18 (18.84)	14 (16.65)
C57BL/6J × F_1	30	3 (10.26)	0 (1.92)	0 (1.25)
C57BL/6J	15	0 (0)	0 (0)	0 (0)
chi square for goodness of fit		11.6 (3 df)	14.5 (3 df)	9.8 (3 df)
probability on a one-locus hypothesis		$p < .01$	$p < .005$	$p < .05$

[a]Numbers in parentheses are expected values under a one-locus, two-allele hypothesis.

hypothesis, and this finding is in good agreement with other reports which indicate that no single measure of audiogenic seizures in these animals can be fit with a single-locus hypothesis.

III. RELATIONSHIP BETWEEN DRUG-, ELECTRICALLY- AND SOUND-INDUCED SEIZURES

We will next review some evidence suggesting a relationship between drug-, electrically-, and sound-induced seizures. Fuller and Wimer (1966) have reviewed some of the pertinent literature and have concluded that in general some relationship exists between susceptibility to audiogenic seizures and susceptibility to chemically- and electrically-induced seizures. Busnel and Lehman (1961), for example, have reported that mice susceptible to audiogenic seizures have lower pentylenetetrazol seizure thresholds than do mice resistant to sound-induced convulsions. Swinyard *et al.* (1963), on the other hand, failed to observe a relationship between susceptibility to sound- and pentylenetetrazol-induced seizures. Although this discrepancy in results could be attributed to any number of variables, it probably occurred because mice of different genotypes were used in these two experiments. Busnel and Lehman (1961) also reported that caffeine produced more seizures in mice susceptible to audiogenic seizures than in mice of resistant genotypes. A number of investigators have reported that electroconvulsive seizure thresholds are lower in mice susceptible to audiogenic seizures than in resistant animals (Goodsell, 1955; Hamburgh and Vicari, 1960; Swinyard *et al.*, 1963). Busnel and Lehman (1961) reported that the ED_{50} for strychnine-induced seizures in mice of a genotype susceptible to audiogenic seizures was 1.35 mg/kg, whereas this threshold was 1.75 mg/kg for mice from a genotype resistant to audiogenic seizures.

Data from our laboratory tend to confirm the general conclusion that DBA/2J mice, which are susceptible to audiogenic seizures, are also more susceptible to other seizure-inducing agents than are C57BL/6J mice which are resistant to audiogenic seizures. Table 2 summarizes results obtained with

TABLE 2. *Pentylenetetrazol and electroconvulsive seizure thresholds in inbred mice*

Seizure	Genotype	Age (days)	N	ED_{50}[a]
pentylenetetrazol	DBA/2J	21	105	31.26
	C57BL/6J	21	105	39.94
	F_1	21	105	35.23
electroconvulsive	DBA/2J	21	40	4.6
	C57BL/6J	21	40	7.0
	F_1	21	40	7.2

[a]For pentylenetetrazol seizures, ED_{50} values are given in mg/kg. These thresholds were calculated using the procedure described by Karman and Spearman (Finney, 1952). For electroconvulsive seizures, ED_{50} values are given as mA of current. These thresholds were obtained by interpolation.

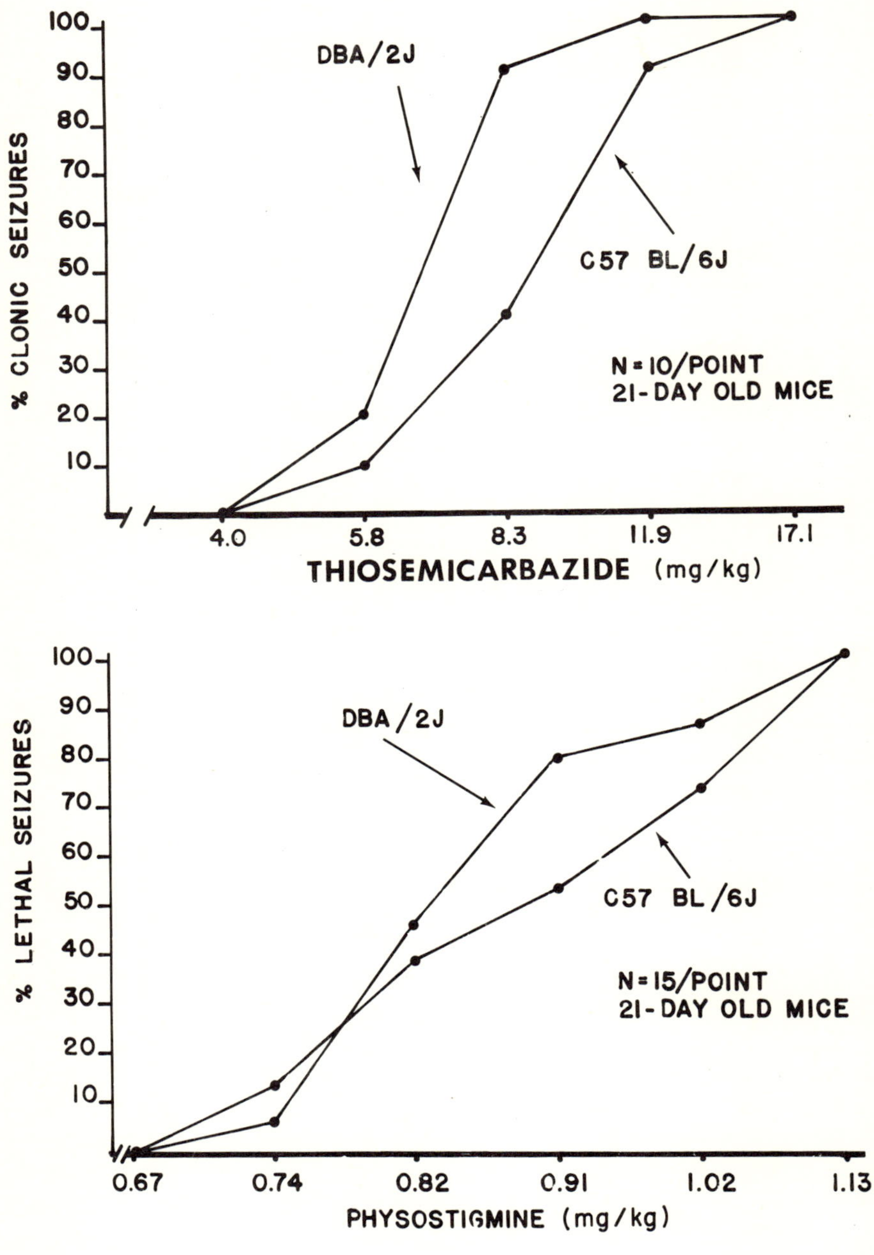

FIG. 2. Effects of genotype on thiosemicarbazide- and physostigmine-induced seizures.

pentylenetetrazol- and electrically-induced seizures in DBA/2J and C57BL/6J mice. These data are taken from larger experiments which we have reported previously (Schlesinger, Boggan, and Griek, 1968). With respect to penthylenetetrazol-induced seizures, DBA/2J mice were found to have slightly lower seizure thresholds than C57BL/6J animals. This difference, although small, was found to be statistically significant. The F_1 hybrid animals were found to be intermediate in susceptibility to pentylenetetrazol-induced seizures. With respect to electroconvulsive seizure thresholds, DBA/2J mice were found to have lower thresholds than both C57BL/6J and F_1 hybrid mice.

Figure 2 illustrates the convulsive effects of physostigmine and thiosemicarbazide in DBA/2J and C57BL/6J mice. DBA/2J mice were found to be slightly more susceptible to physostigmine-induced seizures than were C57BL/6J animals. DBA/2J mice were also observed to be more susceptible to thiosemicarbazide-induced convulsions than were C57BL/6J mice.

We have interpreted these data as preliminary evidence to suggest that DBA/2J mice are more susceptible to seizures than are C57BL/6J mice, regardless of the agents used to induce the convulsions. However, two cautionary remarks must be made. First, the kind of generality implicit in the above statement is very difficult to establish unequivocably since not all convulsants can be tested. Second, even if this type of generality could be established, it would remain an empirical question as to whether or not other genotypes susceptible to audiogenic seizures are also more susceptible to other seizure-inducing agents. Nevertheless, it seems a reasonable assumption that DBA/2J mice could be an excellent animal prototype for studying genetic and biochemical variables which determine seizure susceptibility.

IV. PHENYLALANINE HYDROXYLASE ACTIVITY IN DILUTE AND NON-DILUTE MICE

Coleman (1960) was the first investigator to suggest that susceptibility to audiogenic seizures might be caused by a deficiency in levels of GABA, 5-HT, and NE in brain. Coleman reached this conclusion on the basis of his observations that mice of dilute genotypes, i.e., mice known to be susceptible to audiogenic seizures, were deficient in phenylalanine hydroxylase in liver. The degree of this deficiency varied between dilute genotypes, being most severe in dilute lethal mice and least severe in DBA/2J mice. Rauch and Yost (1963) have also reported that dilute mice have less phenylalanine hydroxylase activity in liver than non-dilute mice. Results from our laboratory, summarized in Table 3, confirm these findings.

It is important to point out that the defect in phenylalanine hydroxylase activity observed in dilute mice is qualitatively different from the enzymatic

TABLE 3. *Phenylalanine hydroxylase activity in 21-day-old C57BL/6J and DBA/2J Mice*

	C57BL/6J		DBA/2J	
	Homogenate[a]	Supernatant	Homogenate	Supernatant
N	12	10	16	10
$\bar{X}$	3.72[b]	3.79[b]	2.59[b]	3.81[b]
S.D.	0.54	0.98	0.96	1.26

[a] Homogenate = crude liver homogenate, homogenized glass to glass. Supernatant = supernatant of 17,000 × *g* for 30 min. The difference in enzyme activity between C57BL/6J and DBA/2J in the crude homogenate is approximately 30% and is statistically significant (t = 3.53, df = 26, $p < 0.01$). The difference in enzyme activity in DBA/2J mice between the crude homogenate and the supernatant fraction is approximately 32% and is statistically significant (t = 2.59, df = 24, $p < .02$).

[b] Enzyme activity was measured by the procedure described by Udenfriend and Cooper (1953) and is expressed as μmoles tyrosine metabolized/g liver/hr.

defect observed in human phenylketonurics. The enzymatic defect in dilute mice does not appear to be a failure to make the enzyme. When the crude liver homogenates are centrifuged at 17,000 × *g*, enzyme activity in the supernatant fraction is the same in dilute and non-dilute mice. This finding has been reported by several investigators, and we have been able to replicate it in our experiments (see Table 3). It has been suggested that in dilute mice an endogenous inhibitor of phenylalanine hydroxylase is associated with the mitochondrial fraction resulting in a decrease in enzyme activity. The nature of this inhibitor remains to be elucidated. In dilute mice the remaining enzyme activity is sufficient to metabolize adequately normal dietary phenylalanine. Experiments in our laboratory have shown that DBA/1J, DBA/2J, C57BL/6J, and C57BL/10J mice have approximately equal levels of serum phenylalanine. Mice of the former two genotypes are examples of dilute mice and mice of the latter two genotypes are examples of non-dilute animals.

Nevertheless, it is possible to demonstrate certain *in vivo* consequences of this partial deficiency in phenylalanine hydroxylase activity in mice of dilute genotypes. For example, if dilute mice are challenged with large amounts of phenylalanine, they metabolize this excess amino acid more slowly than animals of non-dilute genotypes (Coleman, 1960). Coleman has also shown that dilute mice excrete abnormal phenylalanine products even if the animals are maintained on standard laboratory rations. These abnormal phenylalanine metabolites include such compounds as phenylacetic

TABLE 4. *Serum phenylalanine level in inbred mice*[a]

	C57BL/6J	C57BL/10J[b]	DBA/1J	DBA/2J
	14.57	10.71	14.00	14.29
	13.71	11.43	13.71	15.00
	14.29	11.43	13.71	15.29
	14.00	11.72	13.71	14.29
	14.29	11.29	14.00	14.00
X	14.17	11.32	13.83	14.57
S.D.	0.33	0.37	0.16	0.54

[a] All values are expressed as μg phenylalanine/ml serum. Each value was obtained from blood pooled from five mice. All determinations were made in duplicate and average values were used. Duplicate determinations agreed within 10%. All mice were 21 days of age at the time of sacrifice.

[b] Blood from these animals was stored for 5 days. All other determinations were made on fresh blood.

and phenylactic acids. Phenylacetic acid has been shown to inhibit certain decarboxylating reactions. For example, phenylacetic acid inhibits 5-hydroxytryptophan decarboxylase activity in the kidney and also inhibits glutamic acid decarboxylase activity (Hanson, 1958). Sandler and Close (1959) have reported that the production of 5-HT is lowered by phenylacetic acid in certain carcinoid tumor patients. We have performed a number of experiments in which the effects of phenylacetic acid were studied on 5-hydroxytryptophan decarboxylase activity in brain tissue. Using a modification of the procedure described by Snyder and Axelrod (1964), we obtained the results illustrated in Fig. 3. Our results indicate that phenylacetic acid inhibits this enzyme in brain tissue. Figure 3 is a plot of the reciprocal of the substrate concentration against the reciprocal of the reaction velocity (μmole 5-HT formed). Each point on the graph represents the average obtained in from three to six determinations; 20 μmoles of phenylacetic acid was used. These data suggest that phenylacetic acid inhibits the conversion of 5-hydroxytryptophan (5-HTP) to 5-HT in brain tissue and suggests that this inhibition is competitive, although we were unable to prove this point conclusively since at high substrate concentrations 5-HTP itself was observed to inhibit the enzyme.

Since dilute mice produce abnormal phenylalanine metabolites which inhibit decarboxylating reactions, Coleman (1960) suggested that these animals might be deficient in the products of certain reactions in brain tissue. Specifically, Coleman suggested that mice of dilute genotypes might be deficient in levels of GABA, 5-HT, and NE in brain.

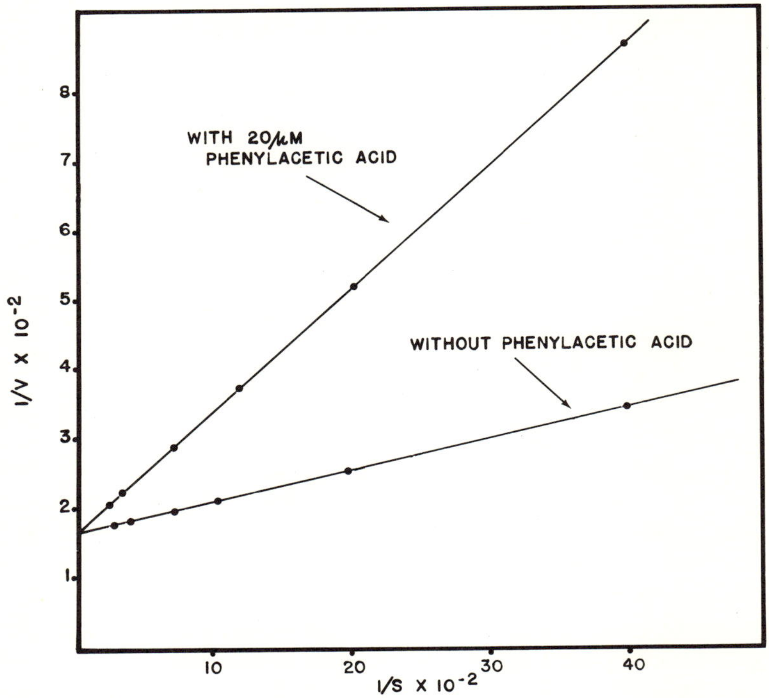

FIG. 3. Effects of phenylacetic acid on 5-hydroxytryptophan decarboxylase activity in brain tissue.

V. STRAIN DIFFERENCES IN LEVELS OF 5-HT AND NE IN BRAIN

Table 5 summarizes the results of two experiments in which levels of 5-HT and NE were measured in brains from DBA/2J and C57BL/6J mice. These data are taken from experiments which we have previously reported (Schlesinger, Boggan, and Freedman, 1965, 1968). In the first experiment, levels of 5-HT and NE determinations were made on whole brain homogenates from 21-day-old DBA/2J and C57BL/6J mice by the spectrofluorometric technique described by Mead and Finger (1961). Levels of both amines were found to be significantly lower in DBA/2J than in C57BL/6J mice. Similar determinations in animals of other ages indicated that differences in levels of these biogenic amines were restricted to this particular developmental period. In the second experiment, levels of 5-HT were determined in the particulate and supernatant fractions from whole brain homogenates from 21-day-old DBA/2J and C57BL/6J mice using the differential centrifugation procedure described by Gilles, Giarman, and Freedman (1964). These

TABLE 5. *Levels of 5-HT and NE in brains of DBA/2J and C57BL/6J mice*

Experiment I	Genotype	N	5-HT (ng/g brain) ± 1 S.D.	NE (ng/g brain) ± 1 S.D.
	DBA/2J	7	466 ± 104	317 ± 73
	C57BL/6J	7	638 ± 28	526 ± 36

Experiment II	Genotype	N	5-HT (ng/whole brain) ± 1 S.D.			
			Particulate	Supernatant	Total	%Supernatant
	DBA/2J	12	212 ± 33	50 ± 11	264 ± 41	19% ± 2.5
	C57BL/6J	12	169 ± 19	40 ± 12	210 ± 23	19% ± 2.5

data are of interest since supernatant 5-HT has been described as "active" and particulate 5-HT as "bound" amine. The data indicate that DBA/2J mice have less 5-HT in both the particulate and supernatant fractions than do C57BL/6J mice; therefore, the ratio of "free" to total 5-HT is not significantly different in animals of these two genotypes.

Certain other data also suggest that DBA/2J mice differ from C57BL/6J mice with respect to the metabolism of these biogenic amines in brain. We have studied the effects of reserpine on depletion of 5-HT and NE in brain; these data are summarized in Fig. 4. In these experiments, 21- and 42-day-old DBA/2J and C57BL/6J mice were injected with vehicle, or with 1.0, 2.0., or 5.0 mg/kg of reserpine. Levels of 5-HT and NE determinations were made 24 hr after drug treatment. The results indicate (1) that reserpine depletes both 5-HT and NE in animals of both genotypes and (2) that reserpine has a significantly smaller effect in older mice and a significantly greater effect in DBA/2J mice (Schlesinger, Boggan, and Freedman, 1968). The mechanisms underlying this genotype by drug treatment interaction remain to be explained. Different rates of drug metabolism by mice of these genotypes could account for these effects.

In order to verify these results, a similar experiment was performed in which we studied the effects of tetrabenazine on 5-HT and NE in brains of DBA/2J and C57BL/6J mice. Tetrabenazine is a synthetic reserpine-like drug which also depletes levels of the biogenic amines in brain tissue. The results of this experiment, summarized in Fig. 5, indicate (1) that this drug depletes levels of 5-HT and NE in animals of both genotypes and (2) that the degree of depletion is greater in DBA/2J than in C57BL/6J mice. This finding is in good agreement with the results obtained with reserpine. One additional finding was obtained in this experiment; namely, rates of repletion of 5-HT and NE are significantly slower in DBA/2J than in C57BL/6J mice.

We have interpreted the results of these experiments as preliminary evidence that DBA/2J mice have lower levels of 5-HT and NE in brain than do C57BL/6J mice, but only at 21 days of age. These data also indicate that there

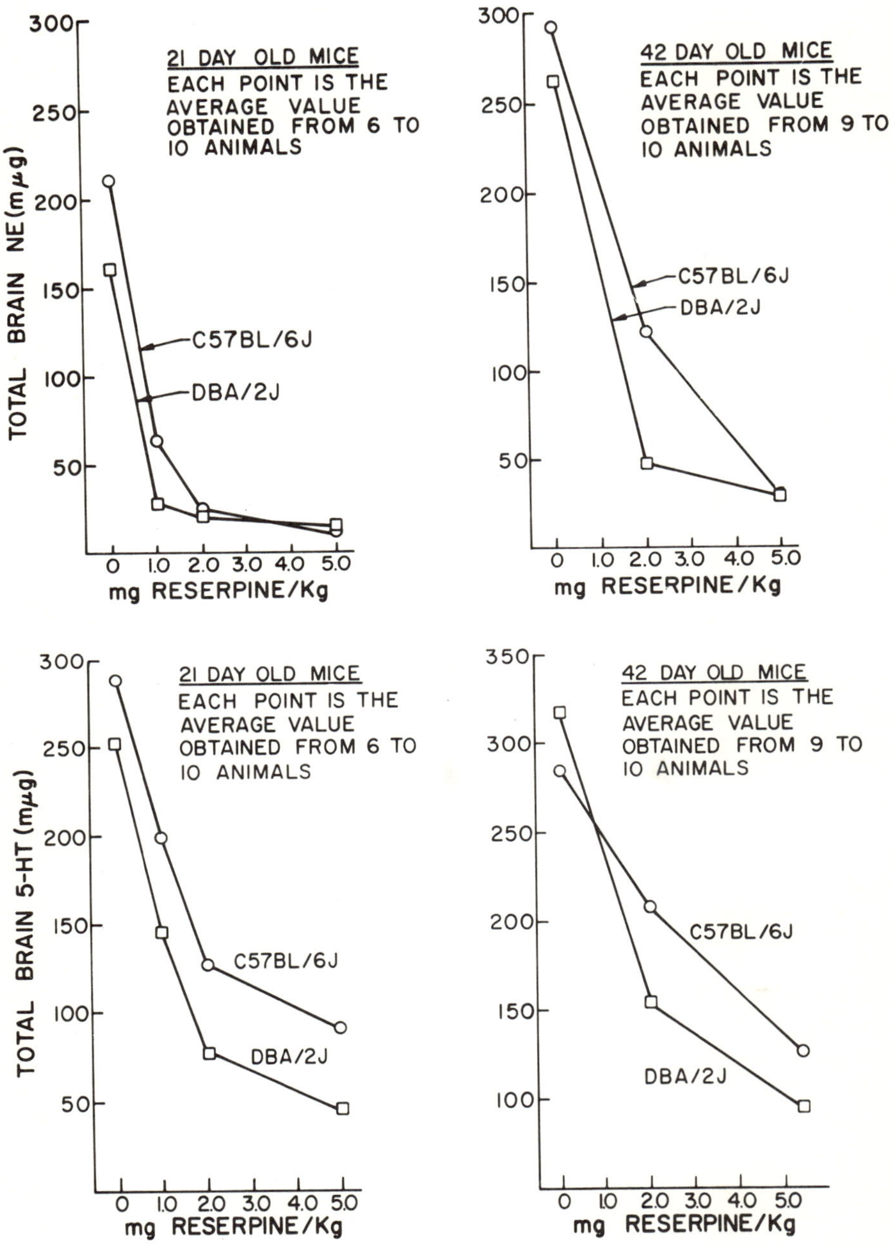

FIG. 4. Effects of reserpine on levels of 5-HT and NE in DBA/2J and C57BL/6J Mice, 24 hr after drug administration.

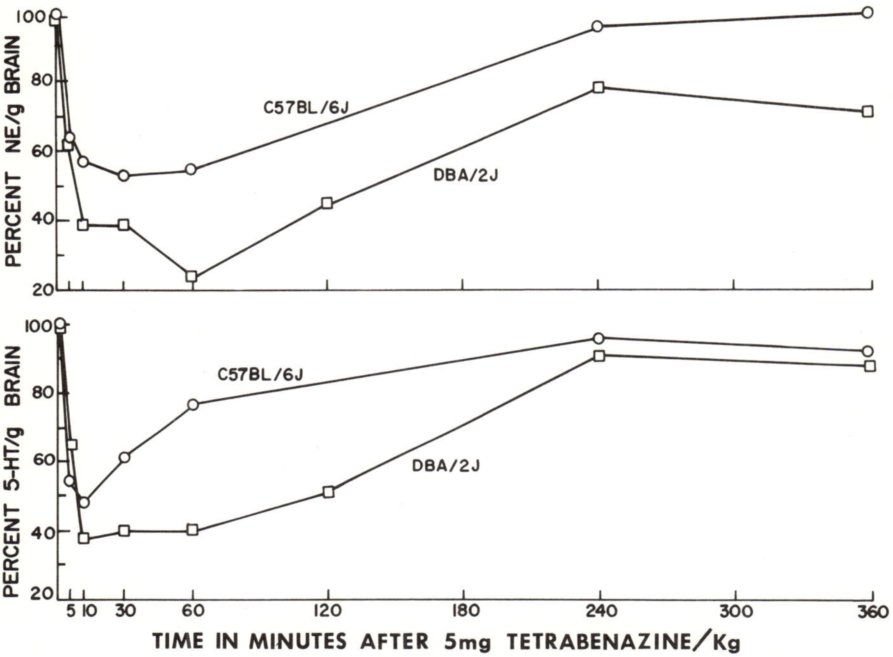

FIG. 5. Effects of tetrabenazine on levels of 5-HT and NE in DBA/2J and C57BL/6J mice as % of 0 time controls.

is a correlation between susceptibility to audiogenic seizures and levels of these biogenic amines in brain. At 21 days of age, DBA/2J mice have lower levels of these compounds in brain and this age corresponds to the period of maximal seizure risk in mice of this genotype. Scudder, Karczmar, Everett, Gibson, and Rifkin (1966) have reported the results of an experiment which adds some support to this hypothesis. These investigators measured levels of biogenic amines in six genera and three inbred strains of mice and determined electroshock-seizure latencies in these animals. They reported that seizure latencies were correlated with levels of the amines, shorter seizure latencies occurring in animals with lower levels of the biogenic amines. Nellhaus (*unpublished data*) has found that rabbits of seizure-prone lines have lower levels of 5-HT in brain than do rabbits of non-susceptible genotypes. On the other hand, McGeer, Ikeda, Asakura, and Wada (1969) failed to find a correlation between levels of amines and susceptibility to audiogenic seizures in DBA/2J and C57BL mice. Their experiment, however, was conducted in 6-week-old animals; in our experiments, we also failed to find differences in levels of 5-HT and NE between DBA/2J and C57BL/6J mice at this age.

VI. PYRIDOXINE METABOLISM AND SUSCEPTIBILITY TO SEIZURES

Lyon, Willams, and Arnold (1958) have reported on the effects of varying amounts of pyridoxine on measures of body weight, food intake, and organ levels of pyridoxine in C57BL/FnLn and I/FnLn mice. These investigators report an unexpected and interesting finding. I/FnLn, but not C57BL/FnLn mice, were made susceptible to audiogenic seizures when maintained on diets deficient in vitamin B_6. I/FnLn mice are homozygous for the autosomal recessive allele *d*, whereas C57BL/FnLn mice are non-dilute. Coleman and Schlesinger (1965) have reported the results of a similar experiment. DBA/2J, BDP/J, and P/J mice (all these strains are dilute), as well as C57BL/6J and C57BL/6J × DBA/2J mice (these mice are non-dilute), were maintained for 3, 5, or 7 weeks on diets containing varying amounts of pyridoxine. Body weight and susceptibility to audiogenic seizures were measured in these mice. With respect to changes in body weight, the following observations were made: DBA/2J mice were severely affected by diets containing zero pyridoxine; these animals gained only 3.5 g after 35 days on this diet. On diets containing somewhat suboptimal levels of pyridoxine (10 mg pyridoxine/kg of diet), these animals gained 6 g over a 35-day period. Animals maintained on diets containing adequate amounts of pyridoxine (30 mg/kg diet) gained 10 g over the 35-day period. The growth curves of the other two dilute strains tested, i.e., BDP/J and P/J mice, were similar to that described for the DBA/2J animals. On the other hand, the growth curves of mice of the non-dilute genotypes, i.e., C57BL/6J and C57BL/6J × DBA/2J, were unaffected by diets containing suboptimal levels of pyridoxine; only diets which contained zero pyridoxine affected growth in animals of these genotypes.

The effects of these diets on susceptibility to audiogenic seizures in mice of the three dilute genotypes used in this experiment are summarized in Table 6. After 3 weeks on their respective diets, mice of all three strains showed an incidence of susceptibility to audiogenic seizures which was inversely proportional to the amount of pyridoxine in the diet. After 5 weeks on these dietary regimens, seizure incidence was also proportional to the degree of pyridoxine deficiency. After 7 weeks on these diets (these data are not given in Table 6), mice of dilute genotypes maintained on the pyridoxine-deficient diets were still susceptible to audiogenic seizures, although an overall decrease in seizure susceptibility was observed. This result is not too surprising since seizure susceptibility is known to decrease with age. C57BL/6J × DBA/2J F_1 hybrid mice which had been maintained on the zero pyridoxine diet for 7 weeks showed some incidence of susceptibility to audiogenic seizures, although the degree of susceptibility was quite low (an 11% incidence of tonic seizures was observed). No seizure episodes were observed in C57BL/6J mice.

TABLE 6. *Effects of pyridoxine-deficient diets on susceptibility to audiogenic seizures in dilute mice*

Genotype	Diet[a]	Seizure incidence (% tonic seizures)	
		3 Weeks	5 Weeks
DBA/2J	1	11% (9)	0% (9)
	2	10% (10)	0% (10)
	3	44% (9)	44% (9)
BDP/J	1	0% (6)	0% (6)
	3	0% (7)	33% (6)
P/J	1	0% (6)	0% (6)
	3	33% (6)	17% (6)

[a] Diets 1, 2 and 3 contained, respectively, 30, 10, and 0 mg of pyridoxine/kg of diet.

[b] Number of weeks refers to the amount of time the animals were maintained on their respective dietary regimes, and the numbers in parentheses refer to the numbers of animals tested.

On the basis of these data and the observations reported by Lyon *et al.* (1958), the conclusion seems warranted that dilute mice are more dependent on dietary pyridoxine than are mice of non-dilute genotypes. In an attempt to replicate and extend the observations reported above, Schlesinger and Schreiber (1969) tested F_2 hybrid mice derived from DBA/2J and C57BL/6J animals for susceptibility to both sound- and electrically-induced seizures following treatment with a pyridoxine-deficient diet. In addition, determinations of levels of 5-HT and NE were made in B_6-deficient and in control animals. Animals were maintained on a zero pyridoxine diet or on a control diet for either 2 or 4 weeks; the animals were also maintained on terramycin-treated (10 g/gallon) tap water in an attempt to make the pyridoxine deficiency more complete. Following these treatments, the animals were tested for either sound- or electrically-induced seizure susceptibility, or were used for 5-HT and NE determinations. The results of these experiments are summarized in Fig. 6. Susceptibility to both sound- and electrically-induced seizures was greater in pyridoxine-deficient than in control animals. With respect to sound-induced seizures, the magnitude of this effect was quite small and certainly not as great as that observed by Coleman and Schlesinger (1965) for audiogenic seizures in mice of dilute genotypes. With respect to electroconvulsive seizures, the increase in susceptibility with pyridoxine deficiency was also quite small. No effects of these treatments on levels of 5-HT and NE in brain were observed.

These data indicate that diets deficient in vitamin B_6 increase susceptibility to audiogenic seizures and to electroconvulsive seizures; the magnitude

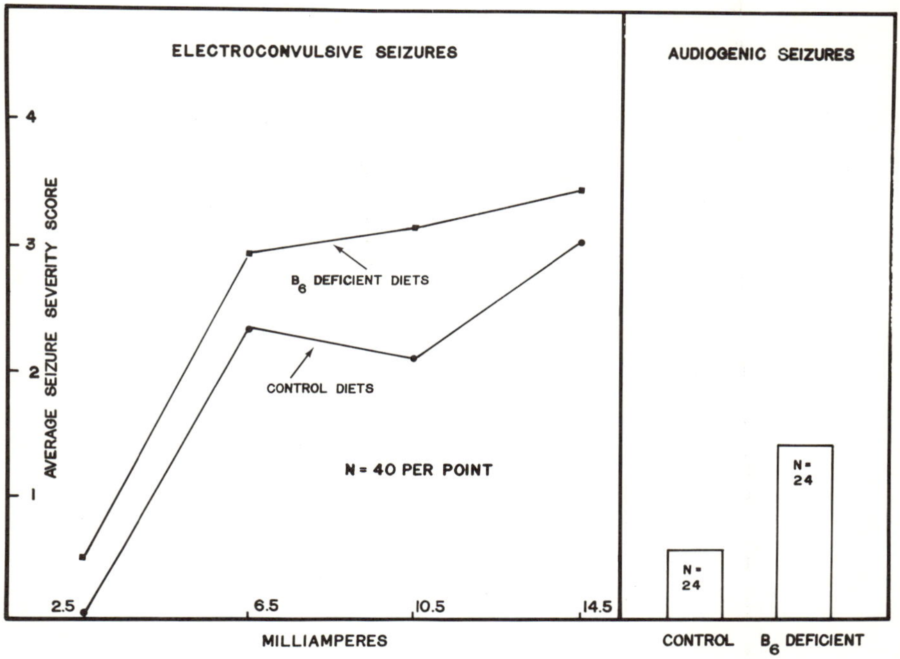

FIG. 6. Effects of pyridoxine deficiencies on susceptibility to electroconvulsive and audiogenic seizures.

of this effect appears to be greater in dilute than in non-dilute mice. If mice of dilute genotypes are indeed pyridoxine dependent, then it would seem reasonable to expect that treatment with supplemental pyridoxine should protect these animals against audiogenic seizures. Schlesinger (1968) has reported the results of a preliminary experiment which bears on this point: from 14 to 21 days of age, DBA/2J mice were given daily injections of 2 mg pyridoxine-HC1 per mouse; control animals were given daily injections of vehicle. Approximately 3 hr after the last injection, at 21 days of age, all animals were tested for susceptibility to audiogenic seizures. The results of these experiments indicate that treatment with supplemental pyridoxine reduces the severity of the seizure episodes. The incidence of lethal seizures in the control group (N = 31) was 74%; the incidence of lethal seizures in the supplemental vitamin B_6 group (N-30) was only 1%. Since this experiment was performed with only a single dose of pyridoxine, and since only a single schedule of treatment was used, these results must be considered as only suggestive; experiments with different treatment schedules and different doses are currently in progress in our laboratory. Nevertheless, it seems very likely that susceptibility to audiogenic seizures can be greatly reduced with vitamin B_6 treatment. These results are very reminiscent of similar data obtained on human patients. Tower (1969) has summarized these data on "familial" pyridoxine dependency, a syndrome which leads to convulsions and which can be controlled by supplemental vitamin B_6.

VII. INTERACTION OF PYRIDOXINE DEFICIENCY AND DRUG TREATMENTS

In a previous section, we have suggested that the susceptibilities to audiogenic seizures observed in mice of certain dilute genotypes might be caused by the lower levels of 5-HT and NE in the brains of these animals. We have also reported the results of certain experiments which suggest that a correlation exists between levels of 5-HT and NE on one hand and susceptibility to seizures on the other. In a later section of this paper, we discuss certain data which indicate that drug treatments which raise or lower levels of these biogenic amines in brain respectively protect against or enhance the action of certain seizure-inducing agents. Nevertheless, the pyridoxine-deficient diets which in our laboratory increase susceptibility to audiogenic seizures do not affect levels of 5-HT and NE in brain. In order to explain the results of these experiments, we have speculated that this increased susceptibility is the result of lowered levels of GABA which are typically associated with pyridoxine-deficient states (Tews and Lovell, 1967; Tews, 1969).

The effects of drug treatments which lower levels of 5-HT and NE and the interactions of such treatments with pyridoxine-deficient states have also been studied. Specifically, F_2 hybrid mice were maintained on either control diets or on a zero pyridoxine diet for 4 weeks; the animals of these two dietary groups were then given one of four drug treatments. The animals were injected with (1) vehicle, (2) α-methyltyrosine (α-MT), (3) para-chlorophenylalanine (*p*-ChPhe), or (4) α-MT *and* *p*-ChPhe. Dose and time parameters were 80 mg/kg α-MT, 2–1/2 hr before seizure tests, and 320 mg/kg *p*-ChPhe, 72 hr before seizure tests. α-MT lowers levels of NE without appreciably altering levels of 5-HT, whereas *p*-ChPhe lowers levels of 5-HT without affecting levels of NE.

The results of these experiments are summarized in Fig. 7. Statistical analyses of these data indicate the following. Analysis of variance of the seizure severity scores indicates that the dietary treatment effect as well as the drug treatment effect were statistically significant ($p < 0.01$). No significant drug × diet interaction was observed. The results of separate t tests indicate the following. (1) Pyridoxine-deficient animals are more susceptible to audiogenic seizures than are mice maintained on control diets. (2) In mice maintained on control diets, no drug treatment further increased susceptibility to audiogenic seizures. (3) Mice maintained on the pyridoxine–deficient diets *and* treated with the various drug regimens were more susceptible to audiogenic

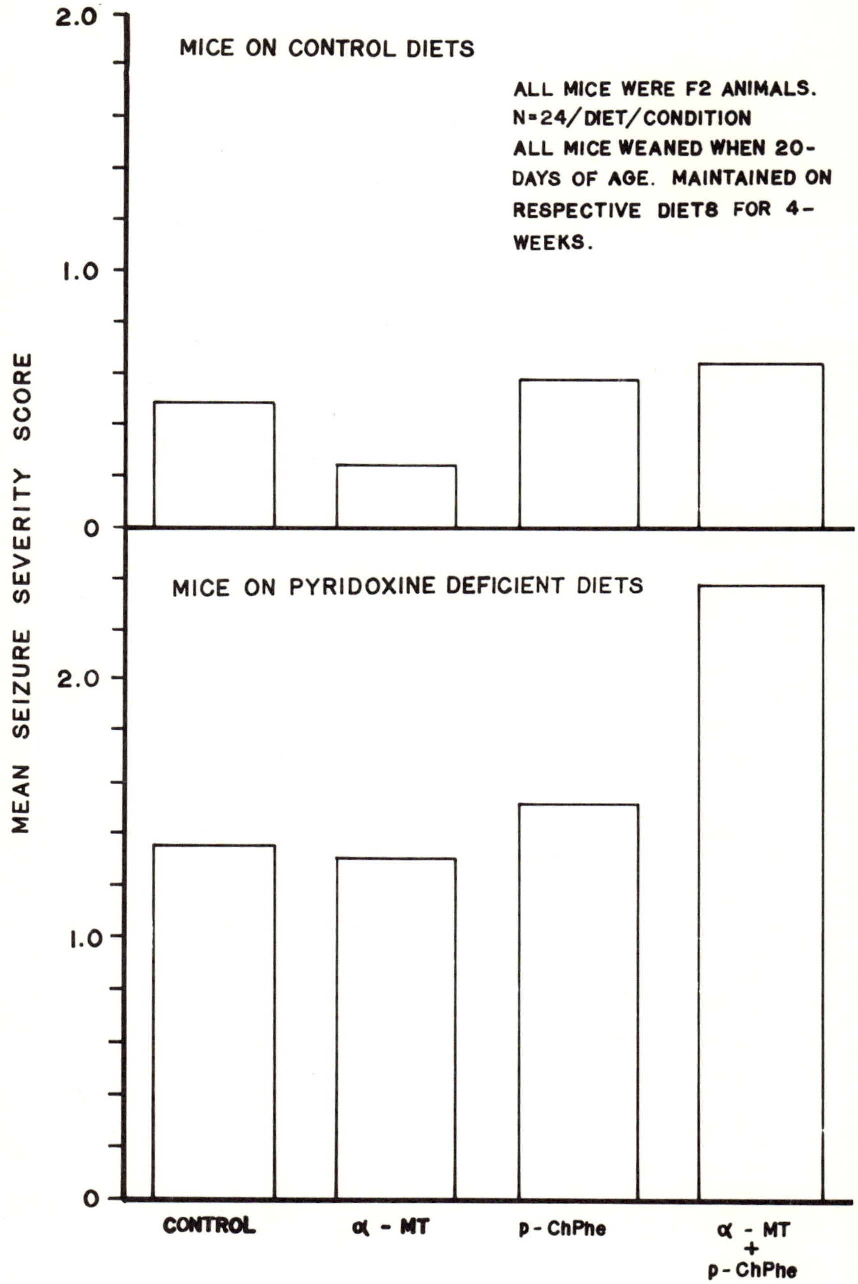

FIG. 7. Effects of α-MT and *p*-ChPhe on susceptibility to audiogenic seizures in control and pyridoxine-deficient mice.

seizures than were the respective control groups. (4) Within the pyridoxine-deficient groups, only treatment with both α-MT *and* *p*-ChPhe further increased seizure susceptibility.

On the basis of these data, we would tentatively conclude that levels of GABA (in this case affected by the dietary regimen) and levels of 5-HT and NE (in this case affected by drug treatment) act synergistically in affecting susceptibility to audiogenic seizures.

VIII. EXPERIMENTALLY-INDUCED PHENYLKETONURIA AND SEIZURE SUSCEPTIBILITY

As we have already suggested, certain similarities exist between human phenylketonurics and dilute seizure-susceptible mice. Both organisms are more susceptible to seizures than control subjects, and human phenylketonurics show an accentuation of both superficial and deep reflexes (Harris, 1962). Nakai, Ida, Kakimoto, Hishikawa, Sano, and Kaneko (1966) have reported that the unaffected heterozygote carrier of phenylketonuria has a lower pentylenetetrazol seizure threshold than does the normal individual. Both organisms show a defect in phenylalanine hydroxylase activity in liver. Because of these and other similarities between these organisms, we have studied the effects of commonly employed techniques for producing experimental phenylketonuria with respect to the effects of these treatments on audiogenic seizures. Specifically, three procedures were used to produce experimental phenylketonuria: (1) treatment with *p*-ChPhe which mimics phenylketogenic effects observed in human patients (Lipton, Gordon, Guroff, and Udenfriend, 1967); (2) diets containing an excess of phenylalanine; and (3) daily injections of phenylalanine. All of these procedures were used in animals of both dilute and non-dilute coat color. The results of these experiments have been reported previously (Schlesinger, Schreiber, Griek, and Henry, 1969).

Table 7 is a summary of the results of an experiment in which the effects of acute administration of *p*-ChPhe and α-MT were studied. Dose and time parameters were: *p*-ChPhe, 320 mg/kg, 72 hr before seizure tests; and α-MT, 80 mg/kg, 2 hr before seizure tests. Four treatment groups were used: a control group injected with vehicle; a group given *p*-ChPhe; a group given α-MT; and a group given both *p*-ChPhe and α-MT. *p*-ChPhe treatment increased susceptibility to audiogenic seizures in both DBA/2J and F_1 hybrid mice; in DBA/2J mice, the magnitude of this effect was small and only marginally significant, whereas in F_1 hybrid mice the effect was highly significant. Similarly, α-MT treatment increased susceptibility to audiogenic seizures in both DBA/2J and F_1 hybrid mice, although in DBA/2J mice this effect was again small and only marginally significant. Treatment with both *p*-ChPhe and α-MT significantly increased seizure susceptibility in both

TABLE 7. *Effects of p-ChPhe and α-MT on susceptibility to audiogenic seizures*[a]

Genotype	Treatment	N	Seizure severity[b] (score ± 1 S.D.)	P^c
DBA/2J	control	46	1.72 ± 1.21	–
	p-ChPhe	44	2.16 ± 1.35	$0.05 < P < 0.1$
	α-MT	19	2.22 ± 1.36	$0.05 < P < 0.1$
	p-ChPhe + α-MT	22	2.36 ± 1.25	< 0.05
F_1	control	11	1.00 ± 0.45	–
	p-ChPhe	10	2.00 ± 1.15	< 0.01
	α-MT	16	1.56 ± 0.73	< 0.01
	p-ChPhe + α-MT	10	3.40 ± 0.31	< 0.01

[a] Similar experiments were performed in C57BL/6J mice. No significant drug effects were observed.

[b] Seizure severity scores: no response = 0; wild running = 1; clonic seizure = 2; tonic seizure = 3; lethal seizure = 4.

[c] Probabilities are based on *t* tests, comparing each drug group against the appropriate control group.

DBA/2J and F_1 hybrid animals. Similar experiments performed in C57BL/6J mice (these data are not included in Table 7) indicate that these treatments do not have any effect in animals of these genotypes.

The effects of chronic administration of phenylalanine on susceptibility to audiogenic seizures in DBA/2J mice are shown in Table 8. Animals were given daily injections of 1.2 mg L-phenylalanine/kg from 14 to 29 days of age. Control animals were given daily injections of vehicle. Another group of animals was treated similarly *and* was given 80 mg α-MT/kg 2 hr before seizure testing. The effects of chronic administration of phenylalanine on levels of 5-HT and NE in brain were also determined. A slight, but not statistically significant, increase in audiogenic seizures was observed after treatment with phenylalanine. Treatment with phenylalanine in combination with α-MT greatly enhanced seizure susceptibility, and this effect was statistically signi-

TABLE 8. *Effects of phenylalanine and α-MT on audiogenic seizures*

Genotype	Treatment	N	Seizure severity[a] (score ± 1 S.D.)	P^b
DBA/2J	control	13	1.30 ± 1.60	–
	phenylalanine	13	1.85 ± 1.28	ns
	phenylalanine + α-MT	10	2.40 ± 0.70	< 0.05

[a] Seizure severity scores: no response = 0; wild running = 1; clonic seizure = 2; tonic seizure = 3; lethal seizure = 4.

[b] Probabilities are based on *t* tests comparing the two treatment groups against the control group. ns = not significant.

TABLE 9. *Effects of diets varying in phenylalanine on audiogenic seizures*

Genotype	Diet[a]	Age[b] (in days)	N	X Seizure severity[c] (score + 1 S.D.)	P^d
DBA/2J	1	7	17	1.53 ± 1.13	
	2		11	2.73 ± 1.42	< 0.05
DBA/2J	1	14	8	1.75 ± 1.28	
	2		10	2.80 ± 0.63	< 0.05
DBA/2J	1	20	17	1.63 ± 1.11	
	2		13	2.54 ± 1.13	< 0.05
C57BL/6J	1	7	10	0	–
	2		5	0.80	–
C57BL/6J	1	14	6	0	–
	2		8	0	–
C57BL/6J	1	20	15	0	–
	2		70	0	–

[a] Diet 1 = control diet; diet 2 = excess phenylalanine.

[b] Age at which the dietary treatments were initiated.

[c] Seizure severity score: no response = 0; wild running = 1; clonic seizure = 2; tonic seizure = 3; lethal seizure = 4.

[d] Probabilities are based on *t* tests comparing the respective dietary groups.

ficant. Chronic administration of phenylalanine lowered levels of 5-HT by approximately 20% without affecting levels of NE in brain.

The effects of dietary manipulations of levels of phenylalanine on susceptibility to audiogenic seizures are illustrated in Table 9. Several dietary regimens were used including control diets and a diet, supplied by Nutritional Biochemical Corp., supplemented with approximately 1% L-phenylalanine by weight (14.6% of the amino acid composition). Weaned 20-day-old DBA/2J and C57BL/6J mice were placed directly on these diets. The mothers of 7- and 14-day-old mice were also placed on these diets until the pups were weaned onto their respective diets at 20 days of age. All seizure tests were conducted at 28 days of age. In DBA/2J mice, excess dietary phenylalanine significantly increased susceptibility to audiogenic seizures. In C57BL/6J mice, a single seizure instance was observed in an animal maintained on the excess phenylalanine diet. Similar experiments, the data for which are not given in Table 9, were performed with mice maintained on diets containing no phenylalanine and no tyrosine. These dietary regimens also increased susceptibility to audiogenic seizures, but only in DBA/2J mice.

These data indicate that phenylketogenic treatments tend to increase susceptibility to audiogenic seizures. Susceptibility is further increased if these treatments are combined with drug treatments which lower levels of NE in the brain. It is especially interesting, however, that DBA/2J and F_1 hybrid mice are more susceptible to these phenylketogenic treatments than are C57BL/6J animals. Since seizures, as we have already indicated, are a common symptom associated with phenylketonuria, it seems likely that mice of these genotypes might serve as a good animal model in which to study this aspect of this disease. The interpretation also seems warranted that treatments which lower levels of 5-HT and NE in brain increase susceptibility to audiogenic seizures.

IX. PHARMACOLOGICAL CORRELATES OF SUSCEPTIBILITY TO SEIZURES

Phenylketogenic treatments, the effects of which include lowering the levels of 5-HT in brain tissue, increase susceptibility to audiogenic seizures. Further, seizure-susceptible mice tend to have lower levels of 5-HT and NE in brain than do mice which are resistant to these seizures. These data strongly suggest that the levels of these biogenic amines are not without significance in determining susceptibility to audiogenic seizures. We have performed a number of pharmacological experiments to study the effects of manipulating levels of these amines and of GABA in terms of the effects of these treatments on seizure susceptibility. The purpose of all these experiments was to determine if the lowering of levels of 5-HT, NE, and GABA in animals of seizure-resistant types made these animals more seizure prone, and, conversely, to determine if increasing levels of 5-HT, NE, and GABA in seizure-susceptible animals protected against seizures. The results of many of these experiments have been reported elsewhere, and only summaries of some of the data will be included here (Schlesinger, Boggan, and Freedman, 1968, 1970; Schlesinger, Boggan, and Griek, 1968).

The effects of reserpine, a drug which lowers levels of 5-HT and NE in the brain, on susceptibility to audiogenic, pentylenetetrazol, and electroconvulsive seizures are summarized in Table 10. In the experiment on audiogenic seizures, 42-day-old DBA/2J and C57BL/6J mice were used. These animals were injected either with 1 mg reserpine/kg or with vehicle, and seizure tests were conducted after 2 hr. The data indicate that reserpine, at this dose, increases susceptibility to audiogenic seizures, but only in DBA/2J mice. Other experiments, with different doses and time parameters, indicate that it is possible to increase susceptibility to sound-induced seizures in C57BL/6J mice, but these effects are always small and never as dramatic as in DBA/2J mice. The numbers of animals used in this experiment were too small for statistical comparisons, but other experiments which utilized larger numbers of mice will be discussed below. For the pentylenetetrazol tests,

TABLE 10. *Effects of reserpine on sound-, pentylenetetrazol- and electrically-induced seizures*

Genotype	Treatment	N	Seizure susceptibility: Number of animals not seizing	Seizure susceptibility: Number of animals seizing[a]	χ^2	*P*
audiogenic seizures						
DBA/2J	control	5	5	0	–	–
	reserpine	5	2	3	–	–
C57BL/6J	control	5	5	0	–	–
	reserpine	5	5	0	–	–
pentylenetetrazol seizures						
DBA/2J	control	10	10	0		
	reserpine	30	1	29	30.47	< 0.001
C57BL/6J	control	10	10	0		
	reserpine	30	0	30	14.40	< 0.001
electroconvulsive seizures						
DBA/2J	control	10	8	2		
	reserpine	10	1	9	7.27	< 0.01
C57BL/6J	control	15	11	4		
	reserpine	15	1	14	11.25	< 0.001

[a] For audiogenic seizures, the occurrence of the clonic-tonic seizures was used as an index of susceptibility. For pentylenetetrazol seizures, the occurrence of muscle spasms, or clonic or tonic seizures was used as an index of susceptibility. For electroconvulsive seizures, the occurrence of the clonic-tonic seizures was used as an index of susceptibility. χ^2 were computed comparing control and reserpine groups.

animals were preinjected either with 2 mg reserpine/kg or with vehicle. After 2 hr, the animals were injected with 32 mg pentylenetetrazol/kg. All animals were 28 days of age. The results indicate that reserpine increases susceptibility to pentylenetetrazol-induced convulsions in both DBA/2J and C57BL/6J mice. For electroconvulsive seizure tests, the mice were preinjected with either 2 mg reserpine/kg or with vehicle. DBA/2J mice were stimulated with 4.6 mA and C57BL/6J mice with 6.9 mA of current. The stimulus was a 833.3 cps square wave of 0.2 sec duration having a negative and positive pulse width of 0.3 msec and a delay of 0.6 msec between each cycle, giving a total period of 1.2 msec. These data indicate that reserpine greatly increases susceptibility to electroconvulsive seizures.

In an attempt to determine the time-response relationships between lowering levels of 5-HT and NE with reserpine and susceptibility to audiogenic seizures, the experiment illustrated in Fig. 8 was performed. Twenty-eight-

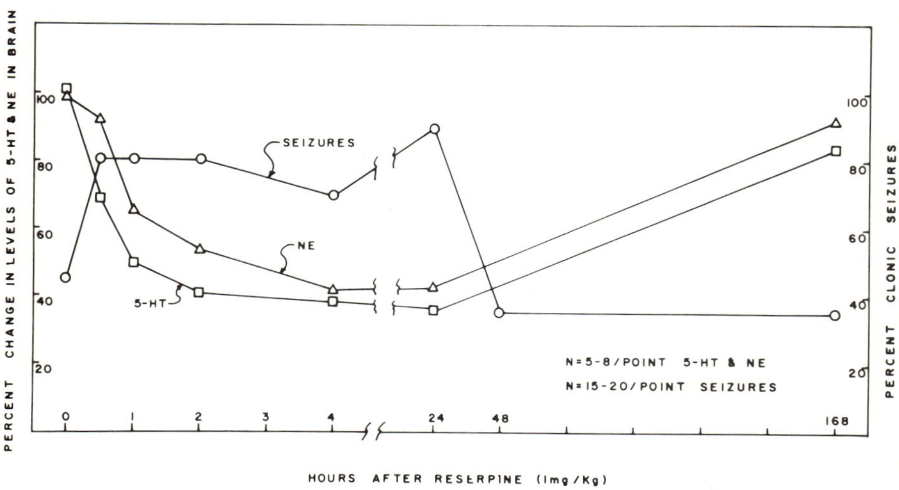

FIG. 8. Effects of reserpine on susceptibility to audiogenic seizures and levels of 5-HT and NE as a function of time.

day-old DBA/2J mice were injected with 1 mg reserpine/kg and tested for susceptibility to audiogenic seizures 0, 1/2, 1, 2, 4, 24, 48, and 168 hr later. Other mice were given similar treatments and determinations of levels of 5-HT and NE were made. Levels of 5-HT and NE decreased slightly 30 min after drug treatment, continued to decrease for 2 hr, remained low for approximately 24 hr, and slowly returned to control values over the next 7-day period. Susceptibility to audiogenic seizures, expressed as the incidence of clonic seizures, increased sharply 30 min after drug treatment, remained high for 24 hr, and then returned to control levels 48 hr after the administration of reserpine.

The results of an experiment in which the effects of administration of 5-HT on susceptibility to sound-, pentylenetetrazol-, and electrically-induced seizures were studied is summarized in Table 11. 5-HTP, the metabolic precursor of 5-HT, raises levels of 5-HT in the brain and should decrease susceptibility to seizures. In the experiment on audiogenic seizures, 21-day-old DBA/2J mice were injected either with 150 mg 5-HTP/kg or with vehicle, and seizure tests were conducted after 3 hr. Pretreatment with 5-HPT significantly decreased the incidence of audiogenic seizures in these animals. For the pentylenetetrazol seizure tests, 28-day-old DBA/2J mice were injected with either 150 mg 5-HTP/kg or vehicle, and 3 hr later were injected with 78 mg pentylenetetrazol/kg. A statistically significant decrease in pentylenetetrazol seizures was observed in these animals after pretreatment with 5-HTP. Similar experiments in C57BL/6J mice indicated that pretreatment with 5-HTP decreased susceptibility to pentylenetetrazol seizures in mice of this genotype. In the electroconvulsive seizure tests,

TABLE 11. *Effects of 5-HTP on sound-, pentylenetetrazol- and electrically-induced seizures*

Genotype	Treatment	N	Seizure susceptibility: Number of animals not seizing	Seizure susceptibility: Number of animals seizing[a]	χ^2	*P*
audiogenic seizures						
DBA/2J	control	16	3	13	8.36	< 0.005
	5-HTP	9	7	2		
pentylenetetrazol seizures						
DBA/2J	Control	21	2	19	5.72	< 0.02
	5-HTP	21	10	11		
electroconvulsive seizures						
DBA/2J	Control	10	3	7	7.91	< 0.01
	5-HTP	10	10	0		

[a] For audiogenic seizures, the occurrence of the clonic-tonic seizures was used as an index of susceptibility. For pentylenetetrazol seizures, the occurrence of muscle spasms, or clonic or tonic seizures was used as an index of susceptibility. For electroconvulsive seizures, the occurrence of the clonic-tonic seizures was used as an index of susceptibility. χ^2 were computed comparing control and 5-HTP groups.

21-day-old DBA/2J mice were injected with either 150 mg 5-HTP/kg or vehicle, and 3 hr later the animals were stimulated with 5.5 mA of current. (The other stimulus parameters were as described for the reserpine experiment.) 5-HTP significantly reduced the incidence of electroconvulsive seizures in animals of this strain. Similar experiments in C57BL/6J mice yielded the same results.

The results of an investigation of the time-response relationship between increasing levels of 5-HT by administration of 5-HTP and susceptibility to audiogenic seizures are summarized in Fig. 9. Twenty-one-day-old DBA/2J mice were injected with 150 mg L-5-HTP/kg and were tested for susceptibility to audiogenic seizures 0, 15, 30, 60, 120, or 240 min later; control animals, i.e., 0 time controls, were injected with vehicle. Similarly treated mice were used for 5-HT determinations. The data indicate that levels of 5-HT increased by approximately 400% within 15 min after treatment with 5-HTP, continued to increase 30 min after injection, and then slowly decreased over the next 4 hr. For 5-HT determinations after 5-HTP treatment, homogenates were washed three times with borate buffer, pH 10. Seizure susceptibility decreased to 0% 15 min after the administration of 5-HTP, was 10% 30 min after drug treatment, and then decreased to control values approximately 2 hr after injection of 5-HTP.

The results of an experiment in which levels of GABA were manipulated

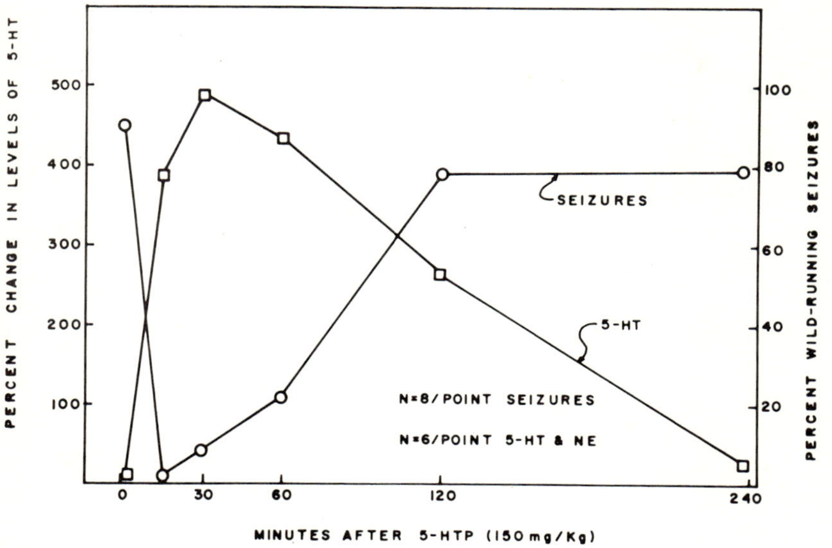

FIG. 9. Effects of 5-HTP on susceptibility to audiogenic seizures and levels of 5-HT as a function of time.

by administration of aminooxyacetic acid (AOAA), and the effects of this treatment on sound-, pentylenetetrazol- and electrically-induced seizures were studied and are summarized in Table 12. AOAA, a compound which increases levels of GABA in brain, would be expected to protect animals against seizures. In an experiment on audiogenic seizures, 21-day-old DBA/2J mice were injected either with 20 μg AOAA/kg or with vehicle, and seizure tests were conducted 7-$\frac{1}{2}$ hr later. The data indicate that pretreatment with AOAA protects these animals against sound-induced convulsions. For pentylenetetrazol seizure tests, 28-day-old DBA/2J mice were injected either with 20 μg AOAA/kg or with vehicle, and 7 hr later with 78 mg pentylenetetrazol/kg. The results indicate that AOAA protects these animals against these seizures. Experiments with C57BL/6J mice indicate that AOAA also protects animals of this genotype against pentylenetetrazol-induced seizures. For electroconvulsive seizure tests, 21-day-old DBA/2J mice were injected either with 20 μg AOAA/g or with vehicle and were stimulated 7 hr later with 5.5 mA of current. (The other stimulus parameters were the same as those described for the reserpine experiment.) AOAA protected these mice against electroconvulsive seizures. Similar experiments with C57BL/6J animals indicated that AOAA also protected mice of this genotype against electroconvulsive seizures.

Figure 10 illustrates the time-response relationship between increasing levels of GABA by AOAA treatment and susceptibility to audiogenic seizures. Animals used in these experiments were injected with 20 μg AOAA/g, and susceptibility to audiogenic seizures was determined 0, 1, 2-$\frac{1}{2}$, 5, 7-$\frac{1}{2}$, 15,

TABLE 12. *Effects of AOAA on sound-, pentylenetetrazol- and electrically-induced seizures*

Genotype	Treatment	N	Seizure susceptibility: Number of animals not seizing	Seizure susceptibility: Number of animals seizing[a]	χ^2	*P*
audiogenic seizures						
DBA/2J	control	10	2	8	13.34	< 0.005
	AOAA	10	10	0		
pentylenetetrazol seizures						
DBA/2J	control	15	1	14	5.71	< 0.02
	AOAA	15	8	7		
electroconvulsive seizures						
DBA/2J	control	10	3	7	5.20	< 0.05
	AOAA	10	9	1		

[a] For audiogenic seizures, the occurrence of the clonic-tonic seizures was used as an index of susceptibility. For pentylenetetrazol seizures, the occurrence of muscle spasms, or clonic or tonic seizures was used as an index of susceptibility. For electroconvulsive seizures, the occurrence of the clonic-tonic seizures was used as an index of susceptibility. χ^2 were computed comparing control and AOAA groups.

and 30 hr after the drug treatment. Determinations of levels of GABA in brain following these treatments are also given; these measurements were kindly provided by Dr. Eugene Roberts, City of Hope Medical Center, Duarte, California. These data indicate that there is a relationship between increasing levels of GABA following treatment with AOAA and decreasing susceptibility to audiogenic seizures.

All of the data described in this section of this paper suggest that levels of 5-HT, NE, and GABA are important determinants of seizure susceptibility, regardless of the agents used to induce the seizures. In general, treatments which lower levels of these amines and of GABA increase susceptibility to seizures, whereas treatments which increase levels of 5-HT, NE, and GABA in brain protect animals against seizures. We have performed other experiments using other drugs and have consistently obtained these types of results. A summary of the effects of these experiments is given in Table 13. In the time-and dose-response experiments we have performed, we have obtained a fairly good correspondence between seizure susceptibility and levels of the biogenic amines and of GABA in brain. The relationship between levels of 5-HT, NE, and GABA and seizure susceptibility, however, is not perfect. A better correspondence between levels of these compounds in brain and seizure susceptibility is obtained when the first derivative, i.e., rate, of levels of the amines is plotted against seizure susceptibility. This suggests that, although

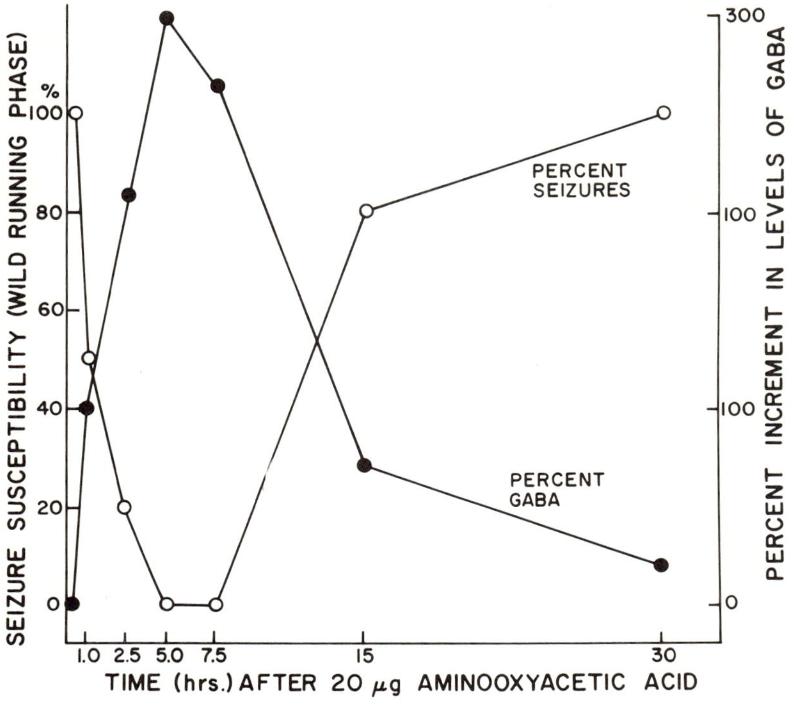

FIG. 10. Effects of AOAA on susceptibility to audiogenic seizures and levels of GABA as a function of time.

TABLE 13. *Summary of the effects of various drugs on seizures*[a]

Drug	5-HT	NE	GABA	Effect on Audiogenic seizures	Effect on Pentylenetetrazol seizures	Effect on Electroconvulsive seizures
reserpine	↓	↓	–	↑	↑	↑
tetrabenazene	↓	↓	–	↑	–	–
α-MT	–	↓	–	↑	–	–
p-ChPhe	↓	–	–	↑	–	–
5-HTP	↑	–	–	↓	↓	↓
iproniazid	↑	↑	–	↓	–	↓
catron	↑	↑	–	↓	–	↓
AOAA	–	–	↑	↓	↓	↓
DOPA	–	↑	–	↓	–	–

[a] Under the 5-HT, NE, and GABA headings, the direction of the arrows have the following meaning: ↑ = increases levels; ↓ = decreases levels; – = no known effect. Under the seizure headings, the direction of the arrows mean the following: ↑ = increases seizure susceptibility; ↓ = decreases seizure susceptibility; – = experiment not carried out.

levels of these compounds *per se* are important determinants of seizure susceptibility, the dynamic states of the 5-HT, NE, and GABA systems are also of importance in determining the response of these organisms to seizure-inducing agents.

The results of these experiments on pharmacological determinants of seizure susceptibility are in good agreement with many reports from other laboratories. Bevan and Chinn (1957), for example, have reported that reserpine increases susceptibility to audiogenic seizures. Kuriyama, Roberts, and Rubinstein (1966) have reported that increasing levels of GABA with AOAA protects mice against convulsions, although the relationship between these variables is quite complex. Ginsburg and Fuller (1954) have reported that the administration of glutamic acid protects mice against seizures. Lehman (1967) has reported that manipulation of levels of the biogenic amines affects seizure susceptibility in the same directions as are obtained in our experiments. Finally, results in our laboratory have shown that direct intracranial administration of 5-HT, NE and/or GABA protects against both sound- and pentylenetetrazol-induced convulsions (Schlesinger, Stavnes, and Boggan, 1969).

XI. SUMMARY AND CONCLUSIONS

The research which we have discussed can be summarized as follows.

1. Certain mice, dilute mice in particular, are more susceptible to audiogenic seizures than are mice of other genotypes.
2. In general, mice susceptible to audiogenic seizure are also more susceptible to other seizure-inducing agents than are mice of genotypes not susceptible to audiogenic seizure.
3. Dilute seizure susceptible mice are deficient in phenylalanine hydroxylase activity in liver.
4. Phenylketogenic treatments, which lower levels of 5-HT, lower seizure thresholds.
5. Dilute seizure-susceptible mice have lower endogenous levels of 5-HT and NE in brain than do mice of seizure-resistant genotypes, but only at 21 days of age–the age of maximal seizure risk in these animals.
6. Treatments which lower levels of 5-HT, NE, and GABA increase susceptibility to seizures, whereas treatments which increase levels of these amines and of GABA protect again seizures.
7. Dilute seizure-susceptible animals may be pyridoxine dependent. In these animals, diets which produce a pyridoxine deficiency increase susceptibility to seizures. In mice of non-dilute genotypes, pyridoxine deficiencies only slightly increase susceptibility to seizures.
8. Pyridoxine deficiencies interact with treatments which lower levels of 5-HT and NE to further enhance seizure susceptibility.

9. Supplemental pyridoxine administration tends to protect animals against audiogenic seizures.

These, then, are some of the data which have been accumulated with respect to susceptibility to seizures in mice of these genotypes. Many questions remain unanswered concerning the mechanisms which determine the inheritance of susceptibility to audiogenic seizures in particular, and to seizures in general. For example, if seizure-susceptible genotypes are indeed deficient in levels of 5-HT and NE in brain, then it seems reasonable to assume that these differences should be localized in certain anatomical areas. The regional distribution of these biogenic amines in seizure-susceptible *versus* seizure-resistant genotypes remains an interesting research problem. It remains unanswered which part of the brain is responsible for initiating audiogenic seizures. Kessner (1966) has shown that decorticate rats are as susceptible to audiogenic seizures as are unoperated controls. Therefore, it seems unlikely that auditory cortex is importantly involved in initiating the seizure response.

Another major problem in audiogenic seizure research concerns the action of 5-HT, NE, and GABA in the central nervous system. Our assumption witn respect to seizures has been that 5-HT, NE, and GABA are inhibitory in their action. Indeed, our data compel us to this conclusion since raising levels of the amines and of GABA protects against seizures, whereas lowering levels of these compounds enhances seizure susceptibility. The issue is not whether these compounds are inhibitory neurohumors; it is sufficient for us to postulate that these compounds are inhibitory in their net action in the central nervous system, or at least in some critical regions. The evidence is fairly conclusive that GABA acts as an inhibitory neurotransmitter. However, with respect to 5-HT and NE, such evidence is largely lacking although Marrazi (1961) has provided some electrophysiological evidence to suggest that 5-HT may have an inhibitory function in the central nervous system. Clearly, much more evidence is necessary to support the view that 5-HT and NE have an inhibitory function *vis-a-vis* seizures.

Although several attempts, in several laboratories, have been made to detect differences in levels of GABA in brains from seizure-susceptible *versus* seizure-resistant mice, none has been successful. Nor, indeed, has it been possible to show differences in the activity of glutamic acid decarboxylase in seizure-susceptible and seizure-resistant mice. If, however, dilute seizure-susceptible mice are pyridoxine dependent as some evidence suggests, then it seems possible that the binding of this vitamin to this enzyme is different in seizure-susceptible and seizure-resistant animals. Commonly used methods for assaying glutamic acid decarboxylase activity add excess co-factor to the reaction mixture, thus obscuring any possible kinetic differences in enzyme activity between mice of these genotypes.

Another important problem is to come to some understanding regarding the genetics of audiogenic seizures. Since single-gene hypotheses do not adequately account for differences in seizure susceptibility between now-existing susceptible and non-susceptible genotypes, a promising research direction might be to look for, or induce, single-gene mutants which are either seizure susceptible or seizure resistant in inbred strains whose phenotypic characteristics with respect to seizures are known. Dr. B. Ginsburg, University of Connecticut, has discovered some such mutants and much can be expected from this type of an approach.

Mice of genotypes not normally susceptible to audiogenic seizures can be made susceptible if the animals are exposed to an intense acoustic stimulus during a critical period in their development. Henry (1967) and Henry and Bowman (1969) were among the first investigators to demonstrate this phenomenon which Henry has termed "priming." Boggan, Freedman, and Schlesinger (1970) have studied the time parameters of this priming effect, and it is of special interest that only a very brief exposure to the priming stimulus is necessary for the effect to appear a day, or even weeks, later. Very little is known about the neurophysiological and biochemical mechanisms which underlie this phenomenon. S. K. Sharpless (*personal communication*) has made the very interesting suggestion that priming may cause some damage to the auditory system and that the priming phenomenon might be due to "supersensitivity of disuse." In order to investigate this phenomenon, the neurophysiological correlates of auditory stimulation must be studied in these animals. With respect to the biochemical mechanisms which underlie the priming phenomenon, some preliminary research in our laboratory has shown that puromycin, given 5 hr before the priming stimulus is applied, abolishes the priming effect.

None of the research we have discussed fully elucidates the causal mechanisms which distinguish seizure-susceptible from seizure-resistant genotypes. It seems obvious that much research needs to be done on this question. However, what seems equally obvious is that an understanding of the mechanisms which mediate the differential sensitivity to seizure-inducing agents in animals of these genotypes will be very useful in understanding some important central nervous system properties.

Acknowledgments. Much of the research discussed in this paper was supported by research grants MH 13026 from the National Institute of Mental Health and GM 14547 from the National Institute of General Medical Sciences. Support from Training Grant MH 11167 is gratefully acknowledged by L. L. Uphouse.

REFERENCES

Bevan, W., and Chinn, R. (1957): Sound-induced convulsions in rats treated with reserpine. *Journal of Comparative and Physiological Psychology, 50*:311.

Boggan, W. O., Freedman, D. X., and Schlesinger, K. (1970): Genetics of audiogenic seizures. IV. Drug induced changes in seizure susceptibility from priming. *Life Sciences* (*in press*).

Busnel, R. G., and Lehmann, A. (1961): Action de convulsivants chemiques sur les souris de lignées sensible et résistante à la crise audiogène. Part III. Cafeine. *Journal of Physiology (Paris), 53*:285.

Chance, M.R.A. (1963): A biological perspective on convulsions. In: *Psychophysiologie Neuropharmacologie et Biochemie de la Crise Audiogène,* edited by R.G. Busnel. Centre National de la Recherche Scientifique, Paris.

Coleman, D. L. (1960): Phenylalanine hydroxylase activity in dilute and nondilute strains of mice. *Archives of Biochemistry and Biophysics, 91*:300.

Coleman, D. L., and Schlesinger, K. (1965): Effects of pyridoxine deficiency on audiogenic seizure susceptibility in inbred mice. *Proceedings of the Society for Experimental Biology and Medicine, 119*:264.

Collins, R. L., and Fuller, J. L. (1969): Audiogenic seizure prone (asp): A gene affecting behavior in linkage group VIII of the mouse. *Science, 162*:1137.

Coursin, D. B. (1969): Vitamin B_6 and brain function in animals and man. *Annals of the New York Academy of Sciences, 166*:7.

Donaldson, H. H. (1924): *The Rat: Data and Reference Tables,* 2nd ed. Wistar Institute, Philadelphia.

Finney, D. J. (1952): *Statistical Method in Biological Assay*. Hafner Publishing Co., Inc., New York.

Frings, H., and Frings, M. (1953): Development of strains of albino mice with predictable susceptibilities to audiogenic seizures. *Science, 117*:283.

Fuller, J. L., and Sjursen, F. H., Jr. (1967): Audiogenic seizures in eleven mouse strains. *Journal of Heredity, 58*:135.

Fuller, J. L., and Thompson, W. R. (1960): *Behavior Genetics.* John Wiley and Sons, Inc., New York.

Fuller, J. L., Wimer, R. E. (1966): Neural, sensory, and motor functions. In: *Biology of the Laboratory Mouse,* edited by E. L. Green. p. 609. McGraw-Hill, New York.

Furuchi, Y. (1969): Clinical and electroencephalographical study of nocturnal epilepsy. *Psychiatry and Neurology Japanica, 71*:101.

Gillis, C. N., Gierman, N. J., and Freedman, D. X. (1964): Retention of 5-hydroxytryptamine by subcellular fractions of rat brain homogenates. *Biochemical Pharmacology, 13*:1457.

Ginsburg, B. E., Fuller, J. L. (1954): A comparison of chemical and mechanical alterations of seizure patterns in mice. *Journal of Comparative and Physiological Psychology*, 47 : 344.

Ginsburg, B. E., and Miller, D. S. (1963): Genetic factors in audiogenic seizures. In: *Psychophysiologie Neuropharmacologie et Biochimie de la Crise Audiogène,* edited by R. G. Busnel. Centre National de la Recherche Scientifique, Paris.

Goodsell, J. S. (1955): Properties of audiogenic seizures in mice and effect of anticonvulsant drugs. *Federation Proceedings, 14*:345.

Hamburgh, M., and Vicari, E. (1960): A study of some physiological mechanisms underlying susceptibility to audiogenic seizures in mice. *Journal of Neuropathology and Experimental Neurology, 19*:461.

Hanson, A. (1958): Inhibition of brain glutamic acid decarboxylase by phenylalanine metabolites. *Naturwissenschaften, 45*:423.

Harris, H. (1959): *Human Biochemical Genetics.* Cambridge University Press, New York.

Henry, K. R. (1967): Audiogenic seizure susceptibility induced in C57BL/6J mice by prior auditory exposure. *Science, 158*:938.

Henry, K. R., and Bowman, R. E. (1969): Effects of acoustic priming on audiogenic, electroconvulsive, and chemoconvulsive seizures. *Journal of Comparative and Physiological Psychology, 67*:401.

Kesner, R. P. (1966): Subcortical mechanisms of audiogenic seizures. *Experimental Neurology, 15*:192.

Kuriyama, K., Roberts, E., and Rubinstein, M. K. (1966): Elevation of gamma-aminobutyric acid in brain with amino-oxyacetic acid and susceptibility to convulsive seizures in mice: A quantitative re-evaluation. *Biochemical Pharmacology, 15*:221.

Langdon-Down, M., and Brain, W. R. (1929): Time of day in relation to convulsions in epilepsy. *Lancet, 1*:1029.

Lehmann, A. (1967): Audiogenic seizures data in mice supporting new theories of biogenic amines mechanisms in the central nervous system. *Life Sciences, 6*:1423.

Lipton, M. A., Gordon, R., Guroff, G., and Udenfriend, S. (1967): p-Chlorophenylalanine-induced chemical manifestations of phenylketonuria in rats. *Science, 156*:248.

Lyon, J. B., Jr., Williams, H. L., and Arnold, E. A. (1958): The pyridoxine-deficient state in two strains of inbred mice. *Journal of Nutrition, 66*:261.

Marrazzi, A. S. (1961): Inhibition as a determinant of synaptic and behavioral patterns. *Annals of the New York Academy of Sciences,* 92:990–1003.

McGeer, E. G., Ikeda, H., Asakura, T., and Wada, J. A. (1969): Lack of abnormality in brain aromatic amines in rats and mice susceptible to audiogenic seizure. *Journal of Neurochemistry, 16*:945.

Mead, J.A.R., and Finger, K. F. (1961): A single extraction method for the determination of both norepinephrine and serotonin in brain. *Biochemical Pharmacology, 6*:52.

Nakai, K., Ida, H., Kakimoto Y., Hishikawa, Y., Sano, I., and Kaneko, Z. (1966): A low metrazol threshold in the heterozygote of phenylketonuria. *Journal of Nervous and Mental Disease, 144*:436.

Nellhaus, G. (1963): Experimental epilepsy in rabbits: Observations on a strain susceptible to audiogenic seizures. In: *Psychophysiologie Neuropharmacologie et Biochimie de la Crise Audiogène,* edited by R. G. Busnel. Centre National de la Recherche Scientifique, Paris.

Partington, M. W. (1961): The early symptoms of phenylketonuria. *Pediatrics, 27*:465.

Rauch, H., and Yost, M. T. (1963): Phenylalanine metabolism in dilute-lethal mice. *Genetics, 48*:1487.

Sandler, M., and Close, H. G. (1959): Biochemical effect of phenylacetic acid in a patient with 5-hydroxytryptophan-secreting carcinoid tumour. *Lancet, 2*:316.

Schlesinger, K. (1968): Experimentally induced seizures in mice. In: *Mind as a Tissue,* edited by Charles Rupp. Harper and Row, New York.

Schlesinger, K., Boggan, W., and Freedman, D. X. (1965): Genetics of audiogenic seizures: I. Relation to brain serotonin and norepinephrine in mice. *Life Sciences, 4*:2345.

Schlesinger, K., Boggan, W., and Freedman, D. X. (1968): Genetics of audiogenic seizures: II. Effects of pharmacological manipulation of brain serotonin, norepinephrine and gamma-aminobutyric acid. *Life Sciences, 7*:437.

Schlesinger, K., Boggan, W. O., and Freedman, D. X. (1970): Genetics of audiogenic seizures: III. Time response relationships between drug administration and seizure susceptibility. *Life Sciences, 9*:721.

Schlesinger, K., Boggan, W., and Griek, B. J. (1968): Pharmacogenetic correlates of pentylenetetrazol and electroconvulsive seizure thresholds in mice. *Psychopharmacologia, 13*:181.

Schlesinger, K., Elston, R. C., and Boggan, W. (1966): The genetics of sound induced seizures in inbred mice. *Genetics, 54*:95.

Schlesinger, K., and Schreiber, R. A. (1969): Interaction of drugs and pyridoxine deficiency on central nervous system excitability. *Annals of the New York Academy of Sciences, 166*:281.

Schlesinger, K., Schreiber, R. A., Griek, B. J., and Henry, K. R. (1969): Effects of experimentally induced phenylketonuria on seizure susceptibility in mice. *Journal of Comparative and Physiological Psychology, 67*:149.

Schlesinger, K., Stavnes, K. L., and Boggan, W. O. (1969): Modification of audiogenic and pentylenetetrazol seizures with gamma-aminobutyric acid, norepinephrine and serotonin. *Psychopharmacologia, 15*:226.

Scudder, C. L., Karczmar, A. G., Everett, G. M., Gibson, J. E., and Rifkin, M. (1966): Brain catecholamines and serotonin levels in various strains and genera of mice and a possible interpretation for the correlations of amine levels with electroshock latency and behavior. *International Journal of Neuropharmacology, 5*:343.

Synder, S. H., and Axelrod, J. (1964): A sensitive assay for 5-hydroxytryptophan decarboxylase. *Biochemical Pharmacology, 13*:805.

Swinyard, E. A., Castellion, A. W., Fink, G. B., and Goodman, L. S. (1963): Some neurophysiological and neuropharmacological characteristics of audiogenic-seizure-susceptible mice. *Journal of Pharmacological and Experimental Therapeutics, 140*:375.

Tews, J. K. (1969): Pyridoxine deficiency and brain amino acids. *Annals of the New York Academy of Sciences, 166*:74.

Tews, J. K., and Lovell, R. A. (1967): The effect of a nutritional pyridoxine deficiency on free amino acids and related substances in mouse brain. *Journal of Neurochemistry, 14*:1.

Tower, D. B. (1969): Neurochemical mechanisms. In: *Basic Mechanisms of the Epilepsies,* edited by H. H. Jasper, A. A. Ward, and A. Pope. Little, Brown and Company, Boston.

Udenfriend, S., and Cooper, J. R. (1953): Assay of L-phenylalanine as phenylethylamine after enzymatic decarboxylation; application to isotopic studies. *Journal of Biological Chemistry, 203*:953.

Vicari, E. M. (1951): Fatal convulsive seizures in the DBA mouse strain. *Journal of Psychology, 33*:79.

Witt, G., and Hall, C. S. (1949): The genetics of audiogenic seizures in the house mouse. *Journal of Comparative and Physiological Psychology, 42*:58.

Advances in Biochemical Psychopharmacology, Vol. 4.
Raven Press, New York © 1972.

Ontogenetic Alterations of Vitamin B_6 Metabolism in Rat Brain—Evidence of Acceleration of its Turnover Rate

Manuchair S. Ebadi

Department of Pharmacology, University of Nebraska College of Medicine, Omaha, Nebraska 68105

Among all coenzymes known to man, vitamin B_6 is the most ubiquitous in nature, reactive in structure, and diversified in function (for reviews see Tower, 1956; Snell, 1958; Braunstein, 1960; Snell, Fasella, Braunstein, and Rossi Fanelli, 1963; Holtz and Palm, 1964; Harris, Wool, and Loraine, 1964; Fasella, 1967; and Snell, this volume). In addition to participating in the metabolism of carbohydrates (see Hedrick, this volume), proteins, lipids, nucleic acids, and coenzymes, vitamin B_6 is involved in the decarboxylation, transamination, racemization, deamination, and desulfhydration of many amino acids (Snell, 1953).

The crucial role that vitamin B_6 plays in the maintenance of chemical integrity of the brain becomes evident when one recalls that all compounds implicated as neurotransmitters—namely, dopamine, norepinephrine, serotonin (Lovenberg, Weissbach, and Udenfriend, 1962), tyramine, tryptamine (Meister, 1965), taurine (Jacobsen and Smith, 1968), histamine (Buffoni, 1966; Kahlson and Rosengran, 1965), gamma-aminobutyric acid (Roberts and Frankel, 1950; Baxter and Roberts, 1958), and even acetylcholine indirectly (Yamada, Sawaki, and Hayami, 1956; Williams and Hata, 1959)—are synthesized and/or metabolized by the aid of the B_6-dependent enzymatic reactions.

Furthermore, the relationships between nutritional deficiencies and development of the central nervous system and mentation (Coursin, 1965), between food deprivation and lack of development of enzyme activity in maturing brain (Swaiman, Daleiden, and Wolfe, 1970), and between hereditary disorders of amino acid metabolism and associated mental deficiency (Berry, 1969), as well as the present belief that memory and learning are related to

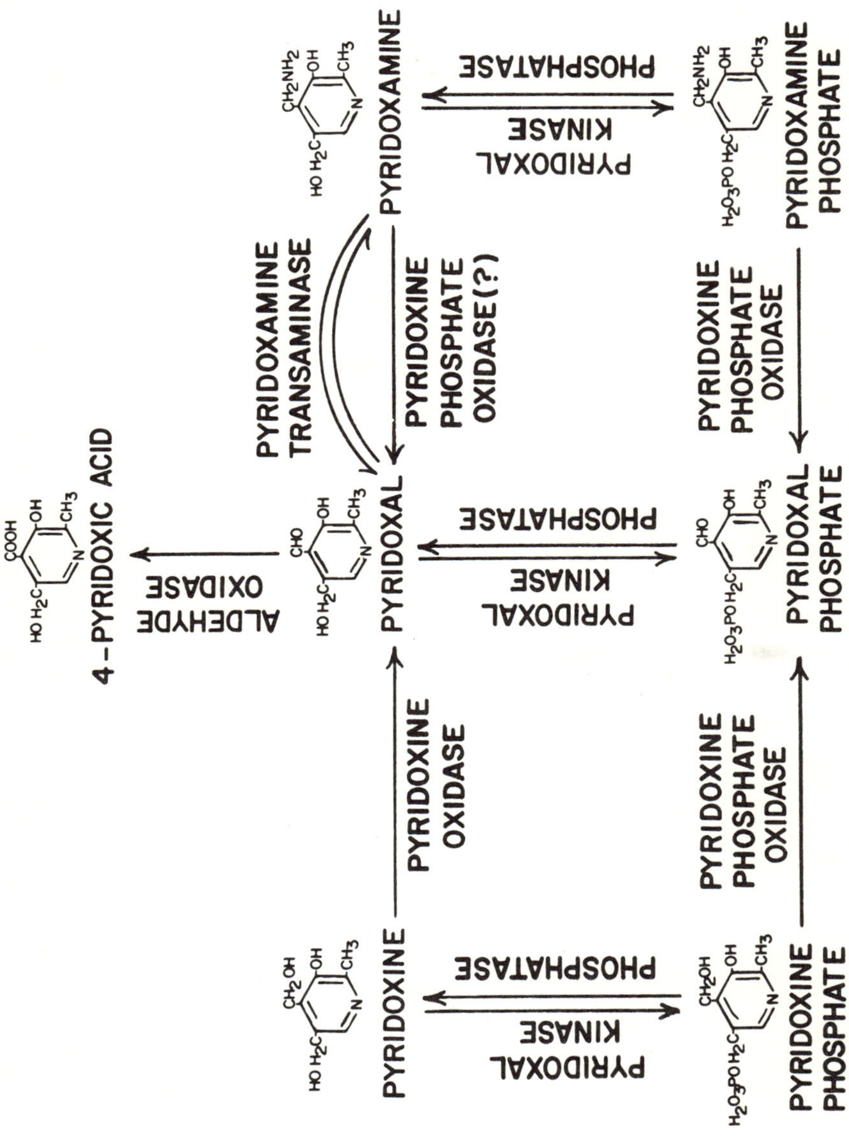

FIG. 1. A summary of the established metabolic transformations of vitamin B_6 in animal tissues (Wada and Snell, 1961).

protein synthesis (Hydén and Lange, 1970), suggest that vitamin B_6 plays crucial role in the developing and mature brain.

In this paper, the alteration in the level of brain pyridoxal phosphate and the enzymes related to its metabolism during development of the brain will be reviewed and discussed. Preliminary evidence that the rate of the synthesis of pyridoxal phosphate increases during development will be introduced.

I. METABOLISM OF VITAMIN B_6

All forms in which B_6 vitamers occur—pyridoxine, pyridoxal and pyridoxamine (Fig. 1)—are phosphorylated *in vivo* by pyridoxal kinase (Hurwitz, 1952, 1953; McCormick, Gregory, and Snell, 1961). However, pyridoxal is the preferred substrate for the mammalian kinases (McCormick *et al.*, 1961). The main pathway for the formation of pyridoxal phosphate is phosphorylation of pyridoxine to pyridoxine phosphate, an important intermediate (Morisue, Morino, Sakamoto, and Ichihara, 1960), and the oxidation of pyridoxine phosphate by pyridoxine phosphate oxidase (Wada, Morisue, Nishimura, Moris, Sakamoto, and Ichihara, 1959; Wada, Morisue, Moris, Sakamoto, and Ichihara, 1960). Pyridoxamine is similarly phosphorylated to pyridoxamine phosphate (Hurwitz, 1953; Podgell, 1958; Wada and Snell, 1961; McCormick *et al.*, 1961) and then readily oxidized to pyridoxal phosphate by the same enzyme that oxidizes pyridoxine phosphate (McCormick *et al.*, 1961). In mammalian brain, only pyridoxal phosphate and pyridoxamine phosphate accumulate; the concentrations of pyridoxine, pyridoxine phosphate, pyridoxal, and pyridoxamine are negligible (Lyon, Bain, and Williams, 1962). At unphysiologically high levels, pyridoxine and pyridoxamine also will be oxidized to pyridoxal; otherwise they are usually metabolized as discussed above.

The phosphorylated coenzymes are degraded by phosphatases (Turner, 1961; Turner and Happold, 1961; Saraswathi and Bachhawat, 1963), and the resulting pyridoxal is converted to 4-pyridoxic acid by the aid of aldehyde oxidase (Huff and Perlzweig, 1944). Recently, Contractor and Shane (1970) have shown that pyridoxal phosphate may be directly oxidized to 4-pyridoxic acid phosphate which in turn may be dephosphorylated to 4-pyridoxic acid. The details of this proposed pathway remain to be confirmed and delineated.

II. ALTERATION IN METABOLISM OF VITAMIN B_6 IN DEVELOPING BRAIN

A. Regional Distribution and the Ontogenetic Diminution in the Level of Pyridoxal Phosphate in the Brain.

The subcellular distribution of B_6 vitamers in rat cerebral cortex has been reported (Loo and Mack, 1971). In studying the regional distribution of pyridoxal phosphate in the rat brain, it was observed that the concentration of pyridoxal phosphate in whole brain declined from 1.30 at birth to 0.75 μg/gram wet tissue in adult animals (Table 1). Consequently, any other studies carried out thereafter have been explorative in nature in an attempt to clarify this observation. In examining the distribution of pyridoxal phosphate in the brain, it becomes clear that the highest concentration of the coenzyme is found in metencephalon, telencephalon and diencephalon in decreasing order, respectively (Ebadi, 1971); and a generalized reduction in coenzyme level between 35–50% occurs everywhere postnatally as the animal matures (Table 1). In order to examine the decline in the level of pyridoxal phosphate more closely, the coenzyme was studied in maturing brain (Fig. 2).

During the first 36 hours after birth, the level of coenzyme showed an apparent elevation, followed by a precipitous drop in 3–5 days, reaching an adult level in approximately 3 weeks (Ebadi, 1970*a*). The decline in the level of pyridoxal phosphate is significant whether calculated on the

TABLE 1. *Regional distribution of pyridoxal phosphate ($\mu g/g$ wet tissue) in rat brain*

Brain region	No. of animals	Newborn	Adult
whole brain	10	1.30 ± 0.16	0.74 ± 0.08
cerebral hemisphere	6	1.15 ± 0.18	0.68 ± 0.12
thalamus	12	0.95 ± 0.10	0.54 ± 0.08
hypothalamus	10	—	0.42 ± 0.05
corpora quadrigemina	10	—	0.35 ± 0.06
pons	8	0.80 ± 0.09	0.45 ± 0.04
cerebellum	8	1.65 ± 0.18	0.90 ± 0.08
medulla oblongata	6	0.74 ± 0.06	0.42 ± 0.06
spinal cord	8	0.88 ± 0.09	0.40 ± 0.05

Male Sprague-Dawley rats were decapitated and various brain regions were separated and kept on aluminum foil, and then frozen together in exactly one minute. This procedure, of course, necessitated the removal of various regions from different rats. Pyridoxal phosphate was separated from deproteinized brain homogenate by a single two-layered ion exchange column chromatography using Dowex 50 potassium and Dowex 1 formate according to the multiphasic continuous gradient elution technique of Bain and Williams (1960). The eluent containing pyridoxal phosphate was concentrated by lyophilization, and its concentration was determined according to the technique of McCormick, Gregory, and Snell (1961).

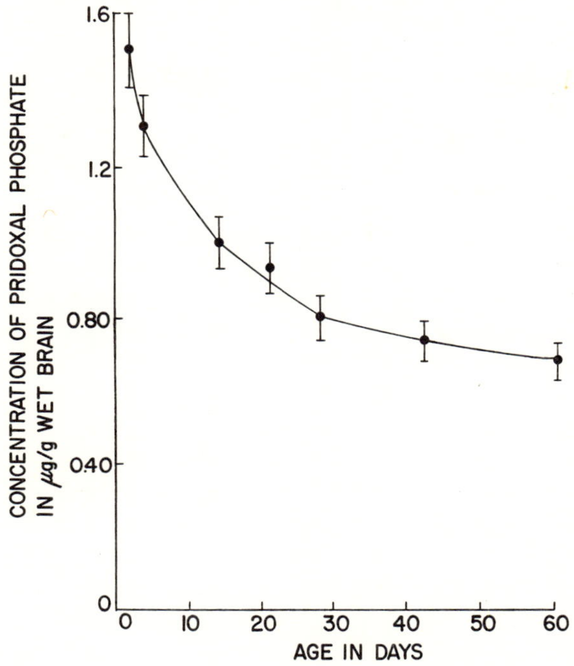

FIG. 2. Ontogenetic alteration in the concentration of pyridoxal phosphate in the brain. Rats were decapitated and the brain separated and frozen on dry ice within 1 min. Pyridoxal phosphate was separated from deproteinized brain homogenate by a single two-layer ion-exchange column chromatography using Dowex 50 potassium and Dowex 1 formate according to the multiphasic continuous gradient elution technique of Bain and Williams (1960). The eluent containing pyridoxal phosphate was concentrated by lyophilization, and its concentration was determined according to the technique of McCormick, Gregory, and Snell (1961).

basis of either dry or wet brains; and whether calculated on either the basis of protein or the amount of DNA. The reduction in the level of pyridoxal phosphate in the rat brain is of extreme academic interest. Similar diminution in the level of pyridoxal phosphate has been reported to occur in leukocytes (Boxer, Pruss, and Goodhart, 1957) and plasma (Hamfelt, 1964) of maturing humans. Hamfelt, by studying the concentration of pyridoxal phosphate in plasma water of human at different ages (1–60), has shown a reduction of approximately five-fold in the level of pyridoxal phosphate. The author eliminated a nutritional defect, defective absorption of vitamin B_6, defective phosphorylation of the vitamin to its phosphorylated active form, and increased loss of the vitamin in the urine, as possible mechanisms for reduction in the level of pyridoxal phosphate. Also, the concentration of pyridoxal phosphate in plasma water of fetal blood is higher than that of the maternal blood (Wachstein, Kellner, and Ortiz, 1960).

Bonasera, Smorto, and Bonavita (1967) have shown a direct relationship between the level of pyridoxal phosphate and isoniazid-induced seizure threshold in developing rat brain. These authors, using the cyanide method of Bonavita (1960), have demonstrated that the level of pyridoxal phosphate declined from 130 to 80 ng/100 mg tissue and that the convulsant dose of isoniazid (given i.p.) correspondingly decreased from 700 to 300 ng/g body weight. This observation is of utmost significance because pyridoxine deficiency causes convulsions in the rat (Chick, ElSadr, and Worden, 1940; Sandman and Snell, 1955) and in the human (Coursin, 1969; see also Schlesinger and Uphouse, *this volume*), and seizure activity and its threshold become altered postnatally (Millichap, 1957; Vernadakis, Curry, Maletta, Irvine, and Timiras, 1966; Vulpe, Caldwell, and Rodin, 1963). Seizure threshold decreases progressively as the animals mature.

The biochemical explanation of the diminution in the brain level of pyridoxal phosphate with age will be discussed under section III. It should be pointed out that the level of another coenzyme, folic acid, decreases 60% in the developing mouse brain (McClain and Bridgers, 1970; see also Bridgers and McClain, this volume). It remains to be studied if the remaining coenzymes demonstrate similar postnatal alteration.

B. Ontogenetic Alteration in the Brain Concentration of 4-Pyriodoxic Acid.

The decline in the level of pyridoxal phosphate prompted us to measure the level of 4-pyridoxic acid, the major end product of vitamin B_6 metabolism. The level of 4-pyridoxic acid declined rapidly from 70 ng/g at birth to 20 ng/g in 4 weeks and maintained a steady state thereafter (Fig. 3; Ebadi, 1970*a*). The decrease in the amount of 4-pyridoxic acid is another indication that the actual available pool of pyridoxal phosphate decreases after birth.

C. Ontogenetic Alteration in the Activity of Pyridoxal Phosphate Phosphatase.

Since the activities of many enzymes increase after birth (see Himwich, 1962, for an extensive review), we decided to study the activity of pyridoxal phosphate phosphatase, the B_6-catabolizing enzyme in the developing brain. The ontogenetic alteration in the activity of this enzyme, with pyridoxal phosphate as a substrate, is shown in Fig. 4. There occurred a precipitous drop in the activity of the enzyme in the first few days after birth, reaching the adult level in approximately 1 week (Ebadi, 1970*a*).

The activity of the nonspecific alkaline phosphatases is high during some stages of morphological differentiation (Chiquoine, 1954) and decreases thereafter. Also, a similar decrease in the level of leukocyte alkaline phosphatase has been shown to occur in humans (O'Kell, 1968). The elevated level of alkaline phosphatase in the fetus is due to elevated production of estrogen (Kaplow, 1955) and progesterone (Brown, 1956) which easily pass

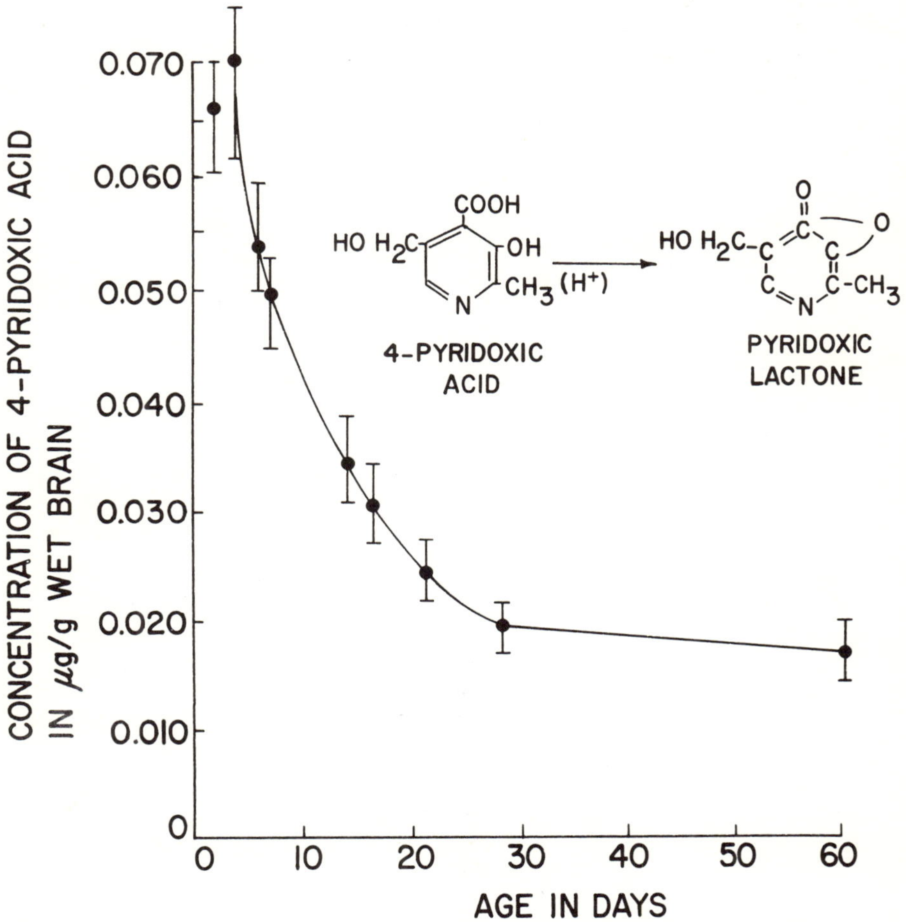

FIG. 3. Ontogenetic alteration in the concentration of 4-pyridoxic acid. Rats were decapitated, and the brain separated and frozen on dry ice in exactly 1 min. 4-Pyridoxic acid was separated from deproteinized brain homogenate by ion-exchange chromatography on Dowex 1 (Cl^-) and Dowex 50 (H^+). The eluted 4-pyridoxic acid was concentrated by lyophilization and converted to pyridoxic acid lactone. The fluorescence of the latter compound was determined spectrophotofluorometrically, essentially according to the method of Reddy, Reynolds, and Price (1958). The 4-pyridoxic acid lactone was indentified according to the method of Ahrens and Korytnyk (1969).

the placental barrier (Avery, 1964). There occurs a precipitous drop in the level of progesterone in the mother immediately prior to delivery (Pritchard, 1957). A concomitant diminution in the level of pregnanediol excretion and alkaline phosphatase activity has been reported to occur in the newborn (Roby, Ober, and Drorbaugh, 1956). Therefore, progesterone is most likely responsible for the rise in the activity of the nonspecific alkaline phosphatase in the fetus.

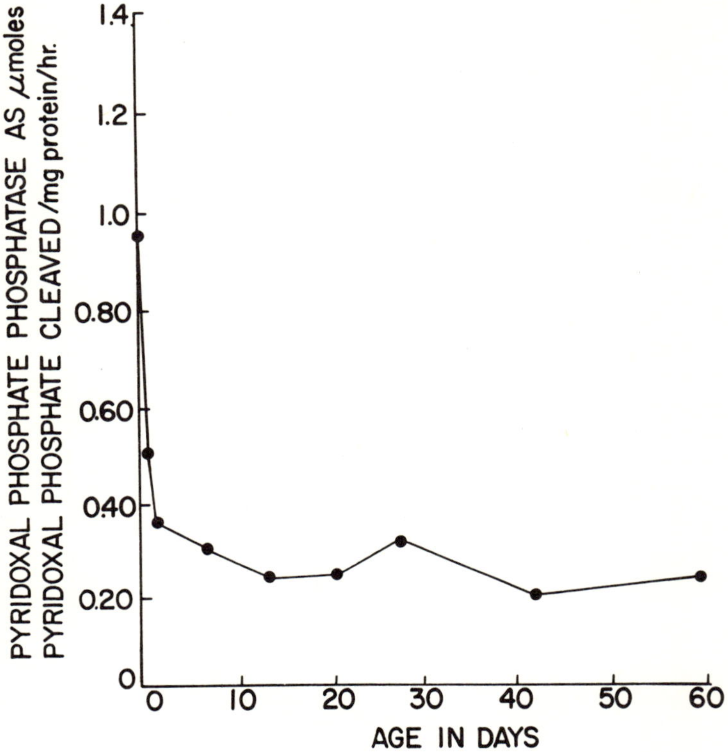

FIG. 4. Ontogenetic alteration in the activity of pyridoxal phosphate phosphatase. The activity of phosphatase, using pyridoxal phosphate as a substrate, was determined by measuring the disappearance of pyridoxal phosphate according to the method of Wada and Snell (1961). The specific activity of the enzyme is reported as μmoles of pyridoxal phosphate hydrolyzed/mg protein/hr.

The diminution in its level heralds the precipitous drop in the activity of this enzyme. Consistent with this hypothesis are the observations that (1) the leukocyte alkaline phosphatase is elevated in women taking synthetic progestational oral contraceptives (O'Kell and Axon, 1965), and (2) the administration of progesterone to male rabbits uniformly resulted in a marked increase in the activity of leukocyte alkaline phosphatase (Ebadi and McCoy, 1966).

Therefore, the rapid drop in the activity of pyridoxal phosphate phosphatase is a nonspecific phenomenon of academic interest and of dubious significance insofar as the overall regulation of coenzyme concentration is concerned.

D. The Development Alteration in the Activities of Pyridoxal Kinase and Pyridoxine Phosphate Oxidase.

Because of the diminishing levels of pyridoxal phosphate and 4-pyridoxic acid, we then turned our attention toward studying the activity of the synthesizing enzymes, pyridoxal kinase and pyridoxine phosphate oxidase. The activity of pyridoxal kinase rose from 0.082 at birth to 0.150 in 4 weeks and to 0.182 μmoles of pyridoxal phosphate formed/mg protein/hr at 8 weeks of age (Fig. 5). Contractor and Shane (1969) have shown a direct correlation between the activity of pyridoxal kinase of human fetal liver and kidney with fetal age. Similarly, the activity of pyridoxine phosphate oxidase became elevated from 1.2 to 4.8 at 4 weeks and 5.2 nmoles of pyridoxal phosphate formed/mg protein/hour at 8 weeks of age (Fig. 6) and maintained a steady state thereafter (Ebadi, 1970*a*).

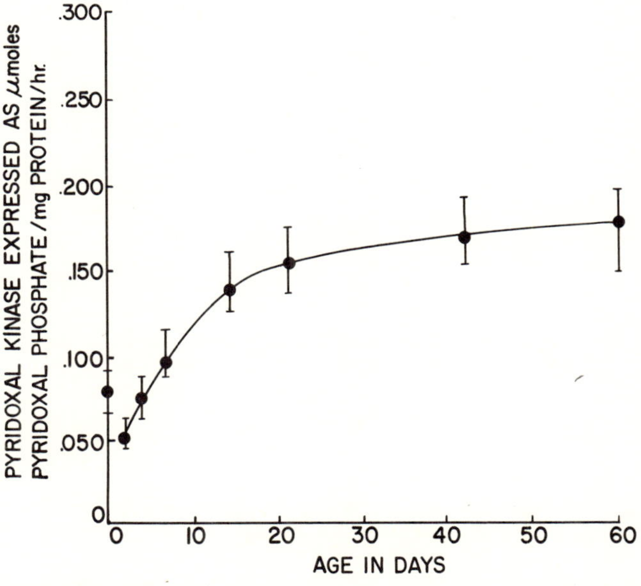

FIG. 5. The developmental alteration in the activity of pyridoxal kinase. Male Sprague-Dawley rats were decapitated, and the brains were immediately removed to beakers in crushed ice. All subsequent operations were carried out at 0-3°C. The cellular particulate fractions of the brain contain materials which mask the activity of the kinase. Thus, the activity of particulate free fractions is considerably higher than the whole-brain homogenate. Therefore, the brain was fractionated according to the differential centrifugation method of Gray and Whittaker (1962). Pyridoxal kinase in the final soluble supernatent fraction was assayed according to the method of McCormick, Gregory, and Snell (1961), using pyridoxal as a substrate and its activity was expressed as μmoles pyridoxal phosphate formed/mg protein/hr. Each value represents the mean of 10 to 15 experiments ± S.E.M.

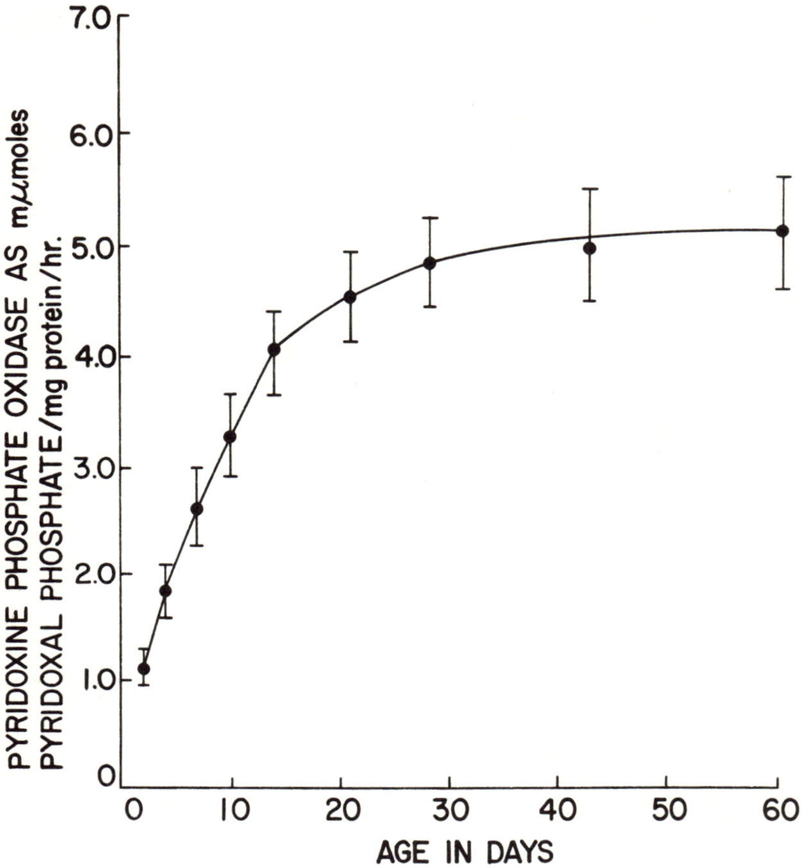

FIG. 6. The developmental alteration in the activity of pyridoxine phosphate oxidase. Male Sprague-Dawley rats were decapitated, and the brains were immediately removed to beakers in crushed ice. The oxidase was partially purified and assayed using pyridoxine phosphate as a substrate, essentially according to the method of Wada and Snell (1961). Its activity was expressed as nmoles pyridoxal phosphate formed/mg protein/hr. Each value represents the mean of 10 to 15 experiments ± S.E.M.

E. Studies with Pyridoxine and Deoxypyridoxine

The studies cited above have shown an inverse relationship between the concentration of pyridoxal phosphate and the activity of pyridoxal kinase. This may be interpreted to indicate that the tissue availability of pyridoxal phosphate regulated the activity of pyridoxal kinase. Although it has been shown indirectly that vitamin B_6 deficiency increases the activity of pyridoxal kinase (Greenberg, Masurat, Rice, Herndon, and VanLoon, 1959), there have been no previous attempts to define the relationship between the activity of pyridoxal kinase and the concentration of pyridoxal phosphate. In an attempt to clarify this interrelationship, the following experiments were undertaken.

Pyridoxine in a dose of 100 mg/kg body weight/day was injected intramuscularly into rabbits for 7 days. A baseline 24-hr urine sample was collected initially for measurement of control excretion of 4-pyridoxic acid, and a final 24-hr sample was collected at the end of the 7-day treatment period. When urine collections for studies of excretion of 4-pyridoxic acid were completed, the rabbits were killed and the brains were removed for the pyridoxal kinase studies.

Deoxypyridoxine in a dose of 100 mg/kg body weight/day was injected intramuscularly into rabbits for 7 days. Before administration of the antimetabolite, the experimental and the control rabbits received 100 mg of L-tryptophan/kg body weight by stomach tube, and a 24-hr urine sample was collected to provide a baseline for excretion of xanthurenic acid in the urine. After the experimental animals had received deoxypyridoxine for 7 days, they were again given the same tryptophan load (100 mg/kg) and the last 24-hr urine sample was collected. The animals were then killed and their brains were removed for pyridoxal kinase studies.

In another series of experiments, 100 mg of pyridoxine/kg/day was administered intramuscularly in three divided doses, and the rabbits were killed at daily intervals for a maximum treatment period of 7 days. In each case, the brain was removed for determination of the level of pyridoxal phosphate and the activity of pyridoxal kinase. A similar set of experiments was conducted with deoxypyridoxine, also administered intramuscularly in a dose of 100 mg/kg/day.

In order to determine if the increase in the activity of pyridoxal kinase was the result of increased synthesis *de novo* of the enzyme, a group of animals were concurrently treated with a single administration of 100 mg of deoxypyridoxine/kg intramuscularly and with doses of 10μg of actinomycin D/kg intravenously repeated every 6 hr, or they were treated with a single administration of 100 mg of deoxypyridoxine/kg intramuscularly and with doses of 10 mg of puromycin/kg intramuscularly repeated every 8 hr. The animals were sacrificed at 2, 4, 6, 12, and 24 hr after initial administration of the drugs.

The daily administration of pyridoxine increased the urinary excretion of 4-pyridoxic acid (Fig. 7), increased the concentration of pyridoxal phosphate in the brain, and decreased the activity of pyridoxal kinase in the brain (Table 2). Under the influence of the B_6-depleting action of deoxypyridoxine, the cerebral concentration of pyridoxal phosphate decreased (Table 3). The administration of deoxypyridoxine promoted an increase in the urinary excretion of xanthurenic acid and in the activity of pyridoxal kinase in the brain (Table 3 and Fig. 8). This increase in the activity of pyridoxal

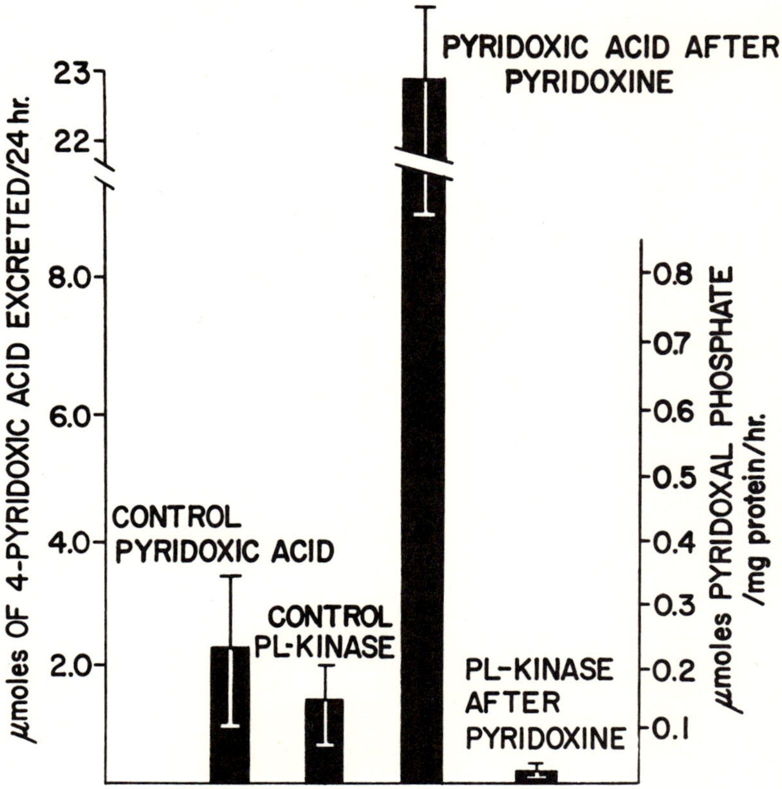

FIG. 7. The effect of administration of pyridoxine on the urinary excretion of 4-pyridoxic acid (expressed and depicted on the left ordinate) and the activity in brain of pyridoxal kinase (expressed and depicted on the right ordinate). 4-Pyridoxic acid was separated from urine by ion-exchange chromatography on Dowex 1 (Cl^-) and Dowex 50 (H^+), and was converted to 4-pyridoxic acid lactone. The fluorescence of the latter compound was determined spectrophotofluorometrically, essentially according to the method of Reddy, Reynolds, and Price (1958). The activity of pyridoxal kinase was determined according to the method of McCormick, Gregory, and Snell (1961). For details, see Ebadi, McCoy, and Kugel (1970).

kinase was blocked by concomitant administration of actinomycin D or puromycin (Table 4; Ebadi, McCoy, and Kugel, 1970).

Injected pyridoxine is metabolized to 4-pyridoxic acid (Huff and Perlzweig, 1944) which accounts for 70 to 90% of the urinary excretion products after administration of any B_6 vitamer (Rabinowitz and Snell, 1949; Linkswiler and Reynolds, 1950), with only 5 to 10% appearing unchanged in the urine (Flexner and Chassin, 1941; Johnson, Hamilton, and Mitchell, 1945). On the other hand, in deficiency of vitamin B_6 there is depression of the activity of kynureninase and kynurenine transaminase (Dalgliesh, Knox, and Neuberger, 1951) with the result that the urinary excretion of kynurenic and xanthurenic acid, normally present in trace amounts (Price

Table 2. *The concentration of pyridoxal phosphate and the activity of pyridoxal kinase in rabbit brain following daily intramuscular administration of pyridoxine (100 mg/kg)*

Day of pyridoxine (PYN) injection	Animal	Pyridoxal phosphate (n moles/g wet wt)	P	Pyridoxal kinase (μmoles/mg protein/hr)	P
1	control	2.63 ± 0.20	<0.08	0.170 ± 0.03	<0.05
	PYN	3.44 ± 0.26		0.130 ± 0.04	
2	control	2.75 ± 0.18	<0.05	0.170 ± 0.02	<0.05
	PYN	3.97 ± 0.31		0.120 ± 0.03	
3	control	2.51 ± 0.26	<0.05	0.170 ± 0.05	<0.05
	PYN	3.93 ± 0.35		0.100 ± 0.04	
4	control	2.80 ± 0.19	<0.05	0.150 ± 0.01	<0.05
	PYN	4.13 ± 0.28		0.070 ± 0.02	
5	control	2.63 ± 0.17	<0.05	0.180 ± 0.13	<0.05
	PYN	4.45 ± 0.28		0.040 ± 0.02	
6	control	2.96 ± 0.20	<0.05	0.170 ± 0.06	<0.05
	PYN	4.66 ± 0.27		0.040 ± 0.02	
7	control	3.16 ± 0.18	<0.05	0.160 ± 0.08	<0.05
	PYN	4.53 ± 0.30		0.030 ± 0.04	

Rabbits were decapitated. Pyridoxal phosphate was separated from deproteinized brain homogenate by a single two-layer ion-exchange column chromatography using Dowex 50 potassium and Dowex 1 formate according to the multiphasic continuous gradient elution technique of Bain and Williams (1960). The concentration of pyridoxal phosphate and the activity of pyridoxal kinase were determined according to the technique of McCormick, Gregory, and Snell (1961). Each group represents an average for five rabbits, and each mean value (± S.E.M.) is compared to its corresponding control. The seven groups of animals received one to seven injections. For details, see Ebadi, McCoy, and Kugel (1970).

Dodge, 1956), becomes copious (Porter, Clark, and Sibler, 1948; Lepkovsky, Roboz, and Haagen-Smit, 1943), such that the level of xanthurenic acid excretion following tryptophan loading has been used as a test for vitamin B_6 deficiency (Leitch and Hepburn, 1961; Wachstein and Gudaitis, 1952*a*, *b*).

The results of the present experiments are interpreted to indicate that in brain the concentration of pyridoxal phosphate controls the activity of the kinase. When pyridoxal phosphate is low, the kinase is elevated, and conversely when pyridoxal phosphate is elevated, the kinase is depressed. Since deoxypyridoxine itself is phosphorylated (Umbreit and Waddell, 1949), it would be difficult to determine to what extent the elevation in the kinase activity is solely due to depleted coenzyme in the brain. In order to test the hypothesis that the availability of pyridoxal phosphate regulates the activity of pyridoxal kinase, the studies with pyridoxine and deoxypyridoxine were undertaken. The administration of pyridoxine increased the urinary excretion of 4-pyridoxic acid 10-fold (from 2.1 to 22.6 μmoles/24 hr) and decreased the cerebral activity of pyridoxal kinase 10-fold (from 0.18

TABLE 3. *The concentration of pyridoxal phosphate and the activity of pyridoxal kinase in rabbit brain following daily intramuscular administration of deoxypyridoxine (100 mg/kg)*

Day of deoxy-pyridoxine (D-PYN) injection	Animal	Pyridoxal phosphate (n moles/g wet wt)	P	Pyridoxal kinase (μmoles/mg protein/hr)	P
1	control	3.24 ± 0.28	<0.05	0.17 ± 0.010	<0.05
	D-PYN	0.61 ± 0.050		0.82 ± 0.040	
2	control	3.60 ± 0.35	<0.05	0.19 ± 0.020	<0.05
	D-PYN	0.73 ± 0.080		0.89 ± 0.040	
3	control	3.16 ± 0.28	<0.05	0.17 ± 0.010	<0.05
	D-PYN	0.49 ± 0.040		0.88 ± 0.040	
4	control	3.04 ± 0.38	<0.05	0.19 ± 0.013	<0.05
	D-PYN	0.98 ± 0.070		0.99 ± 0.050	
5	control	2.91 ± 0.21	<0.05	0.17 ± 0.010	<0.05
	D-PYN	1.05 ± 0.12		0.90 ± 0.040	
6	control	2.63 ± 0.60	<0.05	0.17 ± 0.020	<0.05
	D-PYN	0.05 ± 0.09		0.98 ± 0.050	
7	control	2.83 ± 0.51	<0.05	0.17 ± 0.090	<0.05
	D-PYN	0.97 ± 0.04		0.87 ± 0.040	

Rabbits were sacrificed by decapitation. Pyridoxal phosphate was separated from deproteinized brain homogenate by a single two-layer ion-exchange column chromatography using Dowex 50 potassium and Dowex 1 formate according to the multiphasic continuous gradient elution technique of Bain and Williams (1960). The concentration of pyridoxal phosphate and the activity of pyridoxal kinase were determined according to the technique of McCormick, Gregory, and Snell (1961). Each group represents an average for five rabbits, and each mean value (± S.E.M.) is compared to its corresponding control. The seven groups of animals received one to seven injections. For details, see Ebadi, McCoy, and Kugel (1970).

to 0.015 μmoles of pyridoxal phosphate/mg of protein/hr) (Fig. 7). This indicates that saturation of tissue with pyridoxal phosphate results in reduction in activity of the kinase. This suggestion of a causal relationship between the excretion of 4-pyridoxic acid in the urine and the activity of brain pyridoxal kinase must be interpreted with caution, because the brain pool of pyridoxal phosphate represents only a small portion of the bodily stores of vitamin B_6, and the amount of 4-pyridoxic acid excreted represents the overall B_6 catabolic activity of the body. A more accurate relationship can be studied by direct measurement of the turnover of B_6 vitamers and 4-pyridoxic acid levels in the brain.

Subsequent to administration of deoxypyridoxine, the excretion of xanthurenic acid rose some six-fold (from 5.4 to 32.8 μmoles/24 hr) concomitantly with a six-fold rise in the cerebral activity of pyridoxal kinase (from 0.168 to 1.057 μmoles of pyridoxal phosphate/mg protein/hr) (Fig. 8), further indications that cerebral concentrations of pyridoxal phosphate regulate the activity of pyridoxal kinase. The deoxypyridoxine-induced rise

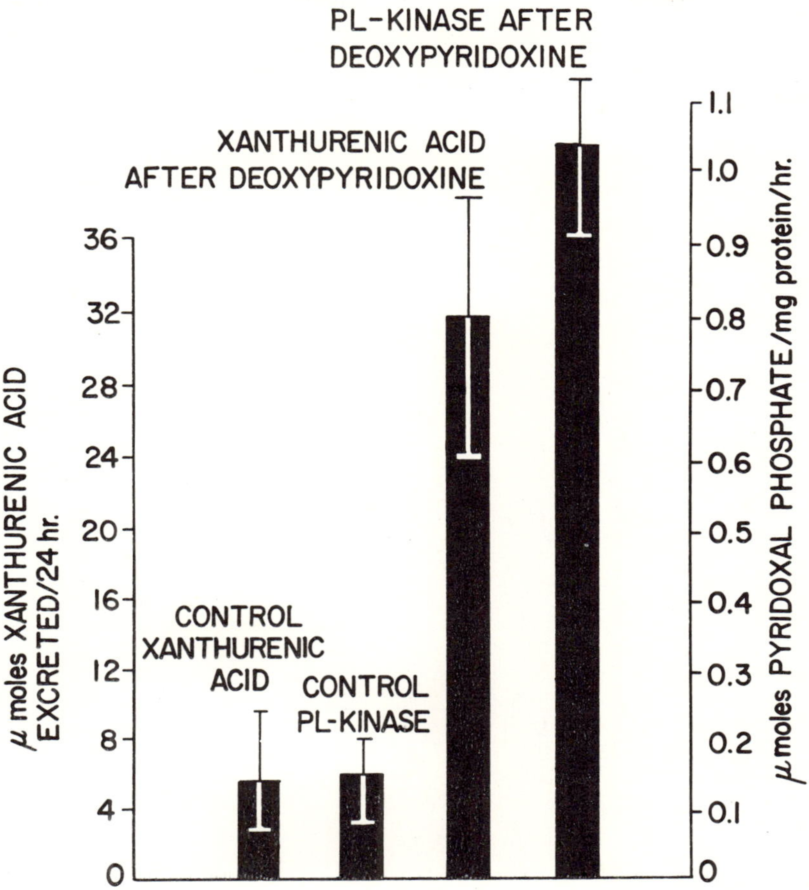

FIG. 8. The effect of administration of deoxypyridoxine on the urinary excretion of xanthurenic acid and the activity in brain of pyridoxal kinase. Xanthurenic acid was isolated by column chromatography using Dowex 50 (H^+) and was rendered fluorescent in strong alkali medium essentially according to the method of Satoh and Price (1958). Pyridoxal kinase was determined according to the method of McCormick, Gregory, and Snell (1961). For details, see Ebadi, McCoy, and Kugel (1970).

in the activity of pyridoxal kinase (observed as early as 4 hr after injection) was blocked by the administration of actinomycin D or puromycin (Table 4), giving evidence that the increase in the activity of pyridoxal kinase may be the consequence of stimulation of activity of messenger or ribosomal RNA.

The foregoing discussion provides evidence that the changes in the activity of pyridoxal kinase may be most directly related to availability of pyridoxal phosphate in the brain. In keeping with this hypothesis are studies which demonstrate increased requirement for B_6 vitamers in high protein diets (Cerecedo, Foy, and DeRenzo, 1948; Beaton, Beare, and McHenry,

TABLE 4. *The effect of administration of actinomycin D (10 μg/kg intravenously every 6 hr) and puromycin (10 mg/kg intramuscularly every 8 hr) on elevation of the activity of pyridoxal kinase in rabbit brain induced by deoxypyridoxine (100 mg/kg)*

Time of killing after initial treatment (hr)	Pyridoxal kinase (μmoles of pyridoxal phosphate/mg of protein/hr					
	Control	Deoxypyridoxine	Deoxypyridoxine and actinomycin D	Deoxypyridoxine and puromycin	Actinomycin D	Puromycin
2	0.18 ± 0.020	0.200 ± 0.010	0.17 ± 0.010	0.17 ± 0.020	0.12 ± 0.020	0.13 ± 0.020
4	0.18 ± 0.020	0.300 ± 0.030	0.16 ± 0.020	0.19 ± 0.020	0.11 ± 0.020	0.12 ± 0.020
			$p < 0.05$	$p < 0.05$		
6	0.19 ± 0.020	0.500 ± 0.030	0.16 ± 0.020	0.20 ± 0.020	0.11 ± 0.010	0.12 ± 0.020
			$p < 0.05$	$p < 0.05$		
12	0.17 ± 0.020	0.780 ± 0.040	0.18 ± 0.030	0.17 ± 0.020	0.09 ± 0.020	0.11 ± 0.010
			$p < 0.05$	$p < 0.05$		
18	0.17 ± 0.010	0.820 ± 0.040	0.20 ± 0.030	0.18 ± 0.030	0.08 ± 0.010	0.11 ± 0.010
			$p < 0.05$	$p < 0.05$		
24	0.18 ± 0.030	0.800 ± 0.040	0.23 ± 0.020	0.17 ± 0.020	0.08 ± 0.010	0.10 ± 0.010
			$p < 0.05$	$p < 0.05$		

Rabbits were decapitated. Pyridoxal kinase activity was determined according to the method of McCormick, Gregory, and Snell (1961). Each group represents an average for five rabbits, and each mean value (± S.E.M.) is compared to its corresponding control. The control animals were injected with the physiological saline (0.154 M NaCl). For details, see Ebadi, McCoy, and Kugel (1970).

1952; Norimitsu, 1959), in hyperthyroidism (Wohl, Levy, Szutka, and Maldia, 1960; Grillo and Pescarmona, 1967), in fever and neoplasm (Wachstein and Lobel, 1956), during pregnancy (Wachstein, 1956; Wachstein and Gudaitis, 1952*a*, *b*; Sprince, Lowy, Folsome, Behrmann, and Raritan, 1951), and in conditions where the brain is replenishing the concentration of the substances which require B_6 for their synthesis (Ebadi, Russell, and McCoy, 1968).

The inverse relationship between the concentration of pyridoxal phosphate and the activity of pyridoxal kinase is not a unique phenomenon. In examining the ontogenetic alteration in the concentration of neurobiologically active compounds, no correlations have been shown to exist between the levels of substrates and by-products, or the activity of the metabolizing enzymes. For example, (1) the activity of Na^+-K^+-activated ATPase in rat brain increases postnatally (Samson and Quinn, 1967), whereas the concentration of ATP remains constant (Lolley, Balfour, and Samson, 1961). (2) The activity of serine transhydroxymethylase in developing brain decreases (Bridgers, 1968), whereas the level of its substrate serine remains constant (Bayer and McMurray, 1967). (3) The levels of norepinephrine and serotonin increase, whereas the activity of aromatic L-amino acid decarboxylase remains constant postnatally (Karki, Kuntzman, and Brodie, 1962). (4) Lastly, the concentration of glutamate and the activity of glutamic acid decarboxylase both increase in the same direction after birth (Himwich, 1962). Therefore, the assessment of the level of substrates and end products might indeed be a poor criterion of biological activity. One must estimate the rate of their synthesis (Costa and Neff, 1968; 1970; see also Costa, this volume).

III. EVIDENCE OF ONTOGENETIC ACCELERATION OF THE RATE OF THE SYNTHESIS AND DEGRADATION OF PYRIDOXAL PHOSPHATE

How does one explain the postnatal reduction in the level of pyridoxal phosphate? One might postulate perhaps that a redistribution and rearrangement of the tissue pool of pyridoxal phosphate may occur. This is highly unlikely, because when one studies the distribution of the vitamin B_6 in the bodies of newborn animals and adults, a similar reduction in the level of pyridoxal phosphate occurs everywhere (Table 5). Also of interest is the similar and negligible level of pyridoxal (< 0.020 $\mu g/g$) in the brains of newborn and adult rats.

A second possibility is that the binding of the pyridoxal phosphate to apoproteins, which exist in phenomenal numbers in the body (e. g., phosphorylase), or to apoenzymes (e.g., alanine racemase, glutamic-aspartic transaminase and glutamic-alanine transaminase) decreases postnatally. The

TABLE 5. *The concentrations of pyridoxal phosphate (μg/g wet tissue or liter plasma) in various tissues of newborn and adult rats*

Tissue	No. of animals	Newborn	Adult
Brain	10	1.3 ± 0.16	0.74 ± 0.88
Liver	12	2.4 ± 0.42	1.18 ± 0.22
Kidney	14	1.8 ± 0.35	0.65 ± 0.42
Muscle	10	2.85 ± 0.70	2.05 ± 0.45
Plasma	75	2.95 ± 0.55	0.95 ± 0.18

Male Sprague-Dawley rats were decapitated and the tissues were removed and processed immediately. Pyridoxal phosphate was separated from deproteinized brain homogenate or plasma by a single two-layer ion-exchange column chromatography using Dowex 50 potassium and Dowex 1 formate according to the multiphasic continuous gradient elution technique of Bain and Williams (1960). Its concentration was determined according to the technique of McCormick, Gregory, and Snell (1961). Plasma from 15 adult or newborn rats was pooled and assayed for pyridoxal phosphate as discussed above.

observation of Adams (1969) shows the following prosthetic binding of pyridoxal phosphate to the above-mentioned enzymes.

Alanine racemase	12.5	μmoles of pyridoxal phosphate/mg protein
Phosphorylase	3.4	"
Glutamic-aspartic transaminase	22.2	"
Glutamic-alanine transaminase	1.46	"

Hence, by decreasing the binding, the necessity of the coenzyme should be anticipated to decrease postnatally. In considering the available evidence, it becomes clear that in the brain the overall rates of protein synthesis (Waelsch and Lajtha, 1961; Schepartz, 1963; Johnson and Luttges, 1966; Schain, Carver, Copenhaver, and Underdahl, 1967) and of nucleic acid synthesis (Barondes, 1964; Mandel, Bieth, and Weill, 1955; Shimizu, Kakimoto, and Sano, 1965; see also Bridgers and McClain, this volume) decrease, whereas the activities of many B_6-dependent enzymes increase after birth (for review, see Himwich, 1962). Furthermore, the actual binding of pyridoxal phosphate by the soluble protein obtained from developing brain increases postnatally (Garten, Singh, and Ebadi, *unpublished data*; Table 6). Also of interest are the studies which show that the uptake

TABLE 6. *The binding of pyridoxal phosphate (μmoles/mg protein) by the soluble brain protein (40,000 g) of the developing rats*

Pyridoxal phosphate added (μmoles)	μmoles of pyridoxal phosphate bound/mg protein						
	Age of the rats (days)						
	Newborn	3	6	9	12	21	60
0.16	0.080	0.12	0.080	0.10	0.14	0.10	0.08
0.32	0.16	0.25	0.18	0.22	0.23	0.20	0.19
0.48	0.25	0.33	0.29	0.36	0.38	0.28	0.25
0.64	0.34	0.43	0.39	0.44	0.51	0.36	0.32
0.80	0.43	0.54	0.51	0.55	0.61	0.48	0.37

Male Sprague-Dawley rats were decapitated, and their brains were homogenized in 0.01 M Tris-HCl buffer, pH 7.4. After centrifugation (40,000 × *g* in a Sorvall Superspeed RC2-B refrigerated centrifuge for an hour), the supernatant was dialyzed against 0.01 M Tris-HCl buffer, pH 7.4, for 24 hr to remove the low molecular weight proteins. The concentration of protein after dialysis was adjusted to 4 mg/ml. A 4-ml quantity of the dialyzed protein was pipetted into cellulose dialyzing bags (16 mm diameter, Union Carbide Corp.) which were placed inside of 25 × 200 mm test tubes containing pyridoxal phosphate ranging from 0.01 to 0.10 μmoles/ml. The tubes were capped with Parafilm to prevent evaporation and covered with aluminum foil to protect the vitamin from decomposition by light. They were then placed horizontally (45°) in a Sorvall Centrifuge tube rack and shaken in an Eberbach shaker at 100 oscillations per min for 45 hr in a cold room (0°–4°C). At equilibrium, pyridoxal phosphate was determined by reacting it with phenylhydrazine according to the method of Wada and Snell (1961). The concentration (μmoles/ml) of pyridoxal phosphate on the outside of the sac was determined and multiplied by the total volume of dialyzing solution. The bound pyridoxal phosphate was obtained by subtracting the unbound value from the total amount initially added at zero time (Ebadi, *unpublished data*).

and binding of other compounds such as noradrenaline in adrenergic nerves (De Champlain, Malmfors, and Olson, 1970) and tubocurarine in cholinergic nerves (Gómez, Pasquini, Soto, and De Robertis, 1970) of rats are enhanced during maturation. Therefore, the maturing brain may increase its ability to bind and store many substances. Consequently, the tentatively proposed decreased binding of pyridoxal phosphate in the brain may not prove compatible with the above-mentioned observations.

The third possibility that might be postulated is that the turnover rate of the synthesis and metabolism of the coenzyme is accelerated postnatally. If this should prove to be the case, it will certainly explain the reduction in the level of coenzyme and the rise in the activity of B_6-synthesizing enzymes along with those of B_6-dependent enzymes.

Borrowing a definition from Costa and Neff (1968; 1970), the term *turnover* refers to the rate of the renewal of a substance in the body or in a tissue. Renewal can be accomplished in different ways: 1) a substance may be synthesized in a given compartment, or 2) a substance may be synthesized elsewhere and reach the compartment via circulation. Since various derivatives of vitamin B_6 are eventually converted to 4-pyridoxic acid (Fig. 1), it was decided to measure the rate of the turnover of the coenzyme by the rate of formation of its end products. Following the assumption that 4-pyridoxic acid efflux from brain, like 5-hydroxyindoleacetic acid, depends on a specialized transport mechanism, the probenecid studies were undertaken. By the technique of Neff, Tozer, and Brodie (1967), 200 mg/kg of probenecid was administered intraperitoneally to rats and the effect on the level of 4-pyridoxic acid in the brain was determined. After 2 hr, only a negligible accumulation of 4-pyridoxic acid was seen (Table 7). Although the idea of measuring the rate of formation of pyridoxal phosphate using this technique is still promising, the mechanism of 4-pyridoxic acid transport from the brain must be delineated and a suitable inhibitor of the said transport system found.

TABLE 7. *The effect of probenecid (200 mg/kg i.p.) on the level of 4-pyridoxic acid in adult rat brain*

Time after treatment (min)	4-Pyridoxic acid (ng/g wet tissue)
15	20 ± 2 (8)
30	22 ± 2 (4)
60	25 ± 3 (4)
90	21 ± 4 (4)
120	32 ± 6 (4)
240	28 ± 4 (4)

Male Sprague-Dawley rats were injected with 200 mg/kg of probenecid i.p. (prepared according to the method of Neff, Tozer, and Brodie, 1967) and then were decapitated at various intervals. 4-Pyridoxic acid was separated from deproteinized brain homogenates by ion-exchange column chromatography on Dowex 1 (Cl^-) and Dowex 50 (H^+). The eluted 4-pyridoxic acid was concentrated by lyophilization and converted to 4-pyridoxic acid lactone. The fluorescence of this latter compound was determined spectrophotofluorometrically, according to the method of Reddy, Reynolds, and Price (1958).

The second approach which proved somewhat fruitful was the study of the rate of accumulation of pyridoxal after inhibition of the aldehyde oxidase by tetraethylthiuramdisulfide (disulfiram) in the brain (Fig. 1). The available pyridoxal in the brain is metabolized to 4-pyridoxic acid according to the following reactions (Hurwitz, 1953):

$$\text{pyridoxal} + \text{aldehyde oxidase} - \text{FAD} \rightarrow \text{pyridoxic acid} + \text{aldehyde oxidase FADH}_2 \quad (1)$$

$$\text{aldehyde oxidase} - \text{FADH}_2 + O_2 \rightarrow \text{aldehyde oxidase FAD} + H_2O_2 \quad (2)$$

The pyridoxal oxidase is competitively blocked by disulfiram (Schwartz and Kzeldgaard, 1951), and this inhibition is reversed by glutathione. Therefore, in a series of experiments, 100 mg/kg of disulfiram was suspended in water and administered orally through a stomach tube to newborn and adult rats. The level of pyridoxal in the brain of 3-day-old animals rose gradually, reached a peak in 12 hr, and declined thereafter (Fig. 9). The level of pyridoxal in the brain of adult rat remained negligible until 6 hr, after which time it rose very rapidly. Similar delay in the onset of action of disulfiram on the oxidase has been reported in rabbit liver (Hald, Jacobsen, and Larsen, 1949). The differences in the rate of accumulation of pyridoxal in the brain of the newborn animal is most probably due to a lower level of oxidase which results in increased susceptibility of the enzyme to inhibition. A lower level of glutathione in the newborn brain (Bayer and McMurray, 1967) may also be responsible for the rapid inactivation of the oxidase. From the data illustrated in Fig. 9, the rate of accumulation of pyridoxal after disulfiram was calculated to be 0.6 nmoles/g/hr in the newborn and 3 nmoles/g/hr in adult animals. Since pyridoxal phosphate is dephosphorylated to and is the immediate precursor of pyridoxal, then the rate of accumulation of pyridoxal may be used as an index of the rate of formation and degradation of pyridoxal phosphate. The rate of formation of 3 nmoles of pyridoxal phosphate/g/hr speaks of a very rapid turnover of the coenzyme in the brain. Indirect evidence gathered by pharmacological intervention (Bain and Williams, 1960; Bilodeau, 1965; Ebadi and Kugel, 1969; Ebadi, 1970*b*) lends credence to this claim.

The disulfiram studies should be examined with care and interpreted with caution. Firstly, the level of pyridoxal in the brain of the newborn and adult rats is below 0.05 μg/g and usually negligible under normal conditions. One must then assume that the regulatory mechanism of brain will prevent the accumulation of this substance. Therefore, after inhibition of aldehyde oxidase, the accumulated pyridoxal may not necessarily correspond to all of the pyridoxal phosphate which has been dephosphorylated. Hence, we may safely assume that some of the accumulated pyridoxal either has effluxed from the brain, or become converted to pyridoxamine or pyridoxal phosphate (Fig. 1). Secondly, since pyridoxal inhibits pyridoxal kinase *in vitro* (White and Dempsy, 1970), the high level of pyridoxal may result in lower formation of pyridoxal phosphate. Therefore, the above-mentioned data at its best

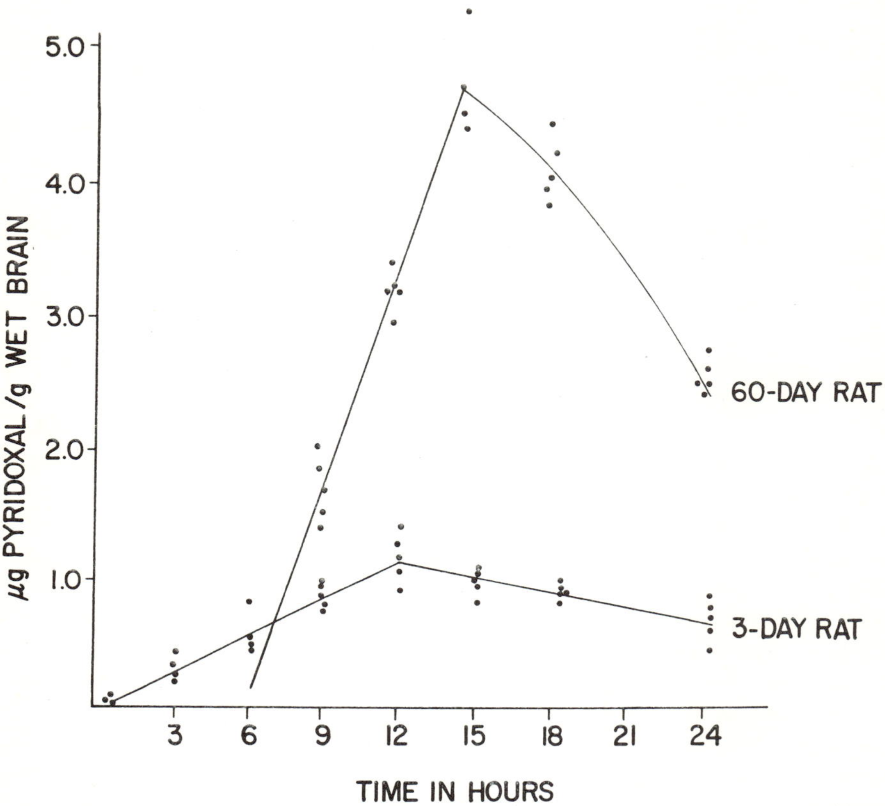

FIG. 9. The effect of tetraethylthiuramdisulfide (disulfiram) on the rate of accumulation of pyridoxal in the brain of newborn and adult rats. Disulfiram was suspended in water and administered by the aid of a stomach tube. The animals were decapitated and the pyridoxal was separated from deproteinized brain homogenate by a single two-layer ion-exchange column chromatography using Dowex 50 potassium and Dowex 1 formated according to the multiphasic continuous gradient elution technique of Bain and Williams (1960). The concentration of pyridoxal was determined according to the method of Bilodeau (1965), and the rate of its accumulation after disulfiram was reported as nmole/g/hr.

provides only a preliminary and qualitative estimation of the rate of biotransformation of pyridoxal phosphate in the brain.

IV. SPECULATION AND CONCLUSION

The term *turnover rate* implies that a steady state exists where the rate of synthesis and degradation of a particular substance in question remains constant (Costa and Neff, 1968, 1970). Examination of the various forms of vitamin B_6 (Fig. 1) reveals that the compounds which are potentially capable of being converted to pyridoxal phosphate are pyridoxine, pyridoxine phosphate, pyridoxamine, pyridoxamine phosphate, pyridoxal, and perhaps 4-pyridoxic acid phosphate (Contractor and Shane, 1970). The concentrations of pyridoxine, pyridoxine phosphate, pyridoxamine, and pyridoxal in the brain are negligible, whereas substantial amounts of pyridoxamine phosphate accumulate. Absolutely no information is available about the rate of interconversion of these substances to pyridoxal phosphate. Although pyridoxamine phosphate is capable of being oxidized to pyridoxal phosphate *in vitro* (Wada and Snell, 1961), the available evidence does not support one such interconversion in the brain. Working with mice, Lyon, Bain, and Williams (1962) demonstrated that avitaminosis induced by a "B_6-deficient ration" did reduce the concentration of pyridoxal phosphate but not of pyridoxamine. They postulated that the reason for the concentration of pyridoxamine remaining unaltered during a deficiency state is that, since 75% of the pyridoxamine phosphate in the brain is bound to mitochondria, it might not be available for the formation of pyridoxal phosphate by the oxidase which is a soluble enzyme (Wada and Snell, 1961). This indicates that the rate of formation and degradation of pyridoxal phosphate and pyridoxamine phosphate are not similar. Therefore, the dynamic equilibrium of the concentration of various B_6 vitamers in the brain must be established prior to determining the rate of formation of pyridoxal phosphate. Furthermore, the nature and the mechanism of 4-pyridoxic acid formation in and transport from the brain must be carefully worked out. The developmental alteration of prosthetic binding of pyridoxal phosphate to apoprotein and apoenzymes (Fasella, 1967) which modifies the catalytic properties of B_6-dependent enzymatic reactions (Sourkes, 1966; Costa, this volume) must be substantiated. Lastly, the contribution of the liver, muscle, kidney, and other tissues, which possess and store vitamin B_6, to the overall pool size of pyridoxal phosphate in the brain must be established. The above-mentioned information in necessary and must be available in order to carry out a fruitful experiment about the turnover rate of a ubiquitously occurring substance such as vitamin B_6 in a complex system such as the brain. Perhaps working with a primitive and simpler neuronal element may be more logical.

The significance of the reduction in the concentration of pyridoxal phosphate in the developing brain is not clear. The actual size of the pool of pyridoxal phosphate may be related directly or indirectly to the regulation of the activity of many B_6-dependent and non-B_6-dependent enzymes. For example, Berman, Justice, and Hsia (1965) have shown that the low activity of 5-hydroxytryptophan decarboxylase in the kidney of the newborn rat may be elevated to adult level simply by administering vitamin B_6. Conversely, the low level of the said decarboxylase may be prolonged when the mothers were placed on a pyridoxine-deficient diet. Blaschko (1950) reported the disappearance of dopa decarboxylase in liver extracts of rats reared on a diet deficient in pyridoxine. Similar alteration in the activities of tyrosine transaminase, tryptophan transaminase, apocysteine sulfinate decarboxylase, threonine dehydrase, and aromatic L-amino acid decarboxylase have been reported and established (for review, see Ebadi, 1970*b*). Of greatest interest are the studies of Tapia, Pasantes, and Massieu (1970) who measured the concentration of pyridoxal phosphate and the activity of glutamate decarboxylase in brains of three vertebrate species—carp, pigeon, and mouse. There existed a direct relationship between the concentration of the coenzyme and enzyme. Carp had the lowest pyridoxal phosphate and glutamate decarboxylase.

The extremely reactive pyridoxal phosphate undergoes Schiff base formation with many amines and amino acids, namely, histidine, histamine, dopamine, norepinephrine, tryptophan, phenylalanine, phenylethylamine derivatives (Schott and Clark, 1952), and glutamic acid (Tapia and Awapara, 1967). The resulting compounds which are pharmacologically and biochemically inactive have been shown to inhibit, among many, the activity of glutamic acid decarboxylase, gamma-aminobutyric acid-transaminase (Topia and Awapara, 1967), and tyrosine transaminase (Black and Axelrod, 1969) *in vitro* and *in vivo*. The activity of pyridoxal kinase may be regulated by varying the concentration of biogenic amines. A series of experiments by Ebadi, Russell, and McCoy (1968) showed an inverse relationship between the concentration of brain dopamine, norepinephrine, and serotonin and the activity of pyridoxal kinase. When the concentration of the amines was lowered by reserpine, the activity of the kinase rose. Conversely, when the concentrations of amines became elevated either by amino acid precursors or by monoamine oxidase inhibition, the activity of kinase became depressed. These results provide further evidence that the concentration of pyridoxal phosphate may regulate the metabolism of the systems which require it.

Also of interest are the studies which have shown that pyridoxal phosphate inhibits the activities of malate dehydrogenase, aldolase, glyceraldehyde 3-phosphate dehydrogenase, glutamic acid dehydrogenase, and 6-phosphogluconic dehydrogenase (Fasella, 1967; Yost and Harrison, 1971). The biological significance of the inhibition of these and many other similar enzymes remains to be clarified.

In conclusion, the delineation of the significance of the ontogenetic reduction in the level of pyridoxal phosphate and the role that it might play in the maintenance of neurochemical integrity of the developing brain will be revealing and rewarding.

Acknowledgment. Supported in part by grants from the National Institute of Child Health and Human Development, USPHS HD-00370, and the National Institute of Neurological Disease and Stroke, NSPHS NS-08932.

REFERENCES

Adams, E. (1969): Fluorometric determination of pyridoxal phosphate in enzymes. *Analytical Biochemistry, 31*:118.

Ahrens, H., and Korytnyk, W. (1969): Pyridoxine chemistry. XXI. Thin-layer chromatography and thin-layer electrophoresis of compounds in the vitamin B_6 group. *Analytical Biochemistry, 30*:413.

Avery, G. B. (1964): A brief review of placental function. *Clinical Proceedings of Children's Hospital (Wash.), 20*:111.

Bain, J. A., and Williams, H. L. (1960): Concentrations of B_6 vitamers in tissues and tissue fluids. In: *Inhibition in the Nervous System and Gamma-Aminobutyric Acid*, edited by E. Roberts, p. 275. Pergamon Press, New York.

Barondes, S. H. (1964): Studies with an RNA polymerase from brain. *Journal of Neurochemistry, 11*:663.

Baxter, C. F., and Roberts, E. (1958): The γ-aminobutyric acid-γ-ketoglutaric acid transaminase of beef brain. *Journal of Biological Chemistry, 233*:1135.

Bayer, S. M., and McMurray, W. C. (1967): The metabolism of amino acids in developing rat brain. *Journal of Neurochemistry, 14*:695.

Beaton, J. R., Beare, J. L., and McHenry, E. W. (1952): Factors affecting the development of acrodynia in pyridoxine-deficient rats. *Journal of Nutrition, 48*:325.

Berman, J. L., Justice, P., and Hsia, D. (1965): The metabolism of 5-hydroxytryptamine (serotonin) in the newborn. *Journal of Pediatrics, 67*:603.

Berry, H. K. (1969): Hereditary disorders of amino acid metabolism associated with mental deficiency. In: *Vitamin B_6 in Metabolism of the Nervous System*, edited by M. A. Kelsall. *Annals of the New York Academy of Sciences, 166*:66.

Bilodeau, F. (1965): Effects of stimulants of the central nervous system on the pyridoxal phosphate content of the rat brain. *Journal of Neurochemistry, 12*:671.

Black, I. B., and Axelrod, J. (1969): Inhibition of tyrosine transaminase activity by norepinephrine. *Journal of Biological Chemistry, 244*:6124.

Blaschko, H. (1950): Substrate specificity of amino acid decarboxylases. *Biochimica et Biophysica Acta, 4*:130.

Bonasera, N., Smorto, M., and Bonavita, V. (1967): Isoniazid seizures in the developing rat and the content of pyridoxal 5-phosphate in the brain. *Brain Research, 4*:383.

Bonavita, V. (1960): The reaction of pyridoxal 5-phosphate with cyanide and its analytical use. *Archives of Biochemistry and Biophysics, 88*:366.

Boxer, G. E., Pruss, M. F., and Goodhart, R. S. (1957): Pyridoxal-5-phosphoric acid in whole blood and isolated leukocytes of man and animals. *Journal of Nutrition, 63*:623.

Braunstein, A. (1960): Pyridoxal phosphate. In: *The Enzymes*, vol. 2, edited by P. D. Boyer; N. Lardy and K. Myrbäck, p. 113. Academic Press, New York.

Bridgers, W. F. (1968): Serine transhydroxymethylase in developing mouse brain. *Journal of Neurochemistry, 15*:1325.

Brown, J. B. (1956): Urinary excretion of oestrogens during pregnancy, lactation and the re-establishment of menstruation. *Lancet, 270*:704.

Buffoni, F. (1966): Histaminase and related amine oxidases. *Pharmacological Reviews, 18*:1163.

Cerecedo, L. R., Foy, J. R., and DeRenzo, E. C. (1948): Protein intake and vitamin B_6 deficiency in the rat. II. The effect of supplementing a low-protein, pyridoxine-deficient diet with cystine or with methionine. *Archives of Biochemistry and Biophysics, 17*:397.

Chick, H., ElSadr, M. M., and Worden, A. N. (1940): Occurrence of fits of an epilepti-form nature in rats maintained for long periods on a diet deprived of vitamin B_6. *Biochemical Journal, 34*:595.

Chiquoine, A. D. (1954): Distribution of alkaline phosphomonoesterase in the central nervous system of the mouse embryo. *Journal of Comparative Neurology, 100*:415.

Contractor, S. F., and Shane, B. (1969): Pyridoxal kinase in the human placenta and foetus through gestation. *Clinica Chimica Acta, 25*:465.

Contractor, S. F., and Shane, B. (1970): 4-pyridoxic acid 5′-phosphate: A metabolite of pyridoxal in the rat. *Biochemical and Biophysical Research Communications, 39*:1175.

Costa, E., and Neff, N. H. (1968): Importance of turnover measurements to elucidate the function of neuronal monoamines. In: *Topics in Medicinal Chemistry*, vol. 2, p. 65. John Wiley and Sons, Inc., New York.

Costa E., and Neff, N. H. (1970): Estimation of turnover rates to study the metabolic regulation of the steady-state level of neuronal monoamines. In: *Handbook of Neurochemistry*, vol. 4, edited by A. Lajtha, p. 45. Plenum Press, New York.

Coursin, D. B. (1965): Effects of undernutrition on central nervous system function. *Nutrition Reviews, 23*:65.

Coursin, D. B. (1969): Vitamin B_6 and brain function in animals and man. In: *Vitamin B_6 in Metabolism of the Nervous System*, edited by M. A. Kelsall. *Annals of the New York Academy of Sciences, 166*:7.

Dalgliesh, C. E., Knox, W. E., and Neuberger, A. (1951): Intermediary metabolism of tryptophan. *Nature, 168*:20.

DeChamplain, C., Malmfors, I., and Olson, R. (1970): The postnatal development of noradrenaline uptake in the adrenergic nerves of different tissues from the rat. *European Journal of Pharmacology, 9*:67.

Ebadi, M. S., and McCoy, E. (1966): Progesterone—mediated increases of leukocyte alkaline phosphatase in rabbits. *Biochemica et Biophysica Acta, 130*:502.

Ebadi, M. S., Russell, R. L., and McCoy, E. E. (1968): The inverse relationship between the activity of pyridoxal kinase and the level of biogenic amines in rabbit brain. *Journal of Neurochemistry, 15*:659.

Ebadi, M. S., and Kugel, R. B. (1969): The opposing effects of amphetamine and chlorpromazine on pyridoxal phosphate concentration and on the activities of pyridoxine phosphate oxidase and pyridoxal kinase in rat brain. *Federation Proceedings, 28*:577.

Ebadi, M. S. (1970*a*): Developmental course of pyridoxal phosphate, pyridoxal kinase, pyridoxal phosphate oxidase, pyridoxal phosphate phosphatase and 4-pyridoxic acid in the CNS. *Federation Proceedings, 29*:1319.

Ebadi, M. S. (1970*b*): Increase in brain pyridoxal phosphate by chlorpromazine. *Pharmacology, 3*:97.

Ebadi, M. S., McCoy, E. E., and Kugel, R. B. (1970): Interrelationships between pyridoxal phosphate and pyridoxal kinase in rabbit brain. *Journal of Neurochemistry, 17*:941.

Ebadi, M. S. (1971): Regional distribution and ontogenetic diminution of the concentration of pyridoxal phosphate in the rat brain. *Federation Proceedings, 30*:382.

Fasella, P. (1967): Pyridoxal phosphate. *Annual Review of Biochemistry, 36*:185.

Flexner, J., and Chassin, M. R. (1941): Clinical studies on pyridoxine (vitamin B_6). *Journal of Clinical Investigation, 20*:313.

Gómez, C. J., Pasquini, J. M., Soto, E. F., and DeRobertis, E. (1970): Binding of D-[dimethyl-^{14}C] tubocurarine, acetylcholinesterase and proteolipids in subcellular membranes of cerebral cortex of the developing rat. *Journal of Neurochemistry, 17*:1485.

Gray, E. G., and Whittaker, V. P. (1962): The isolation of nerve endings from brain: An electron-microscopic study of cell fragments derived by homogenization and centrifugation. *Journal of Anatomy, 96*:79.

Greenberg, S. M., Masurat, T., Rice, E. G., Herndon, J. F., and VanLoon, E. J. (1959): The effect of chlorpromazine hydrochloride on glutamic acid decarboxylase and pyridoxal kinase in rat brain homogenates. *Biochemical Pharmacology, 2*:308.

Grillo, M. A., and Pescarmona, G. P. (1967): Metabolism of pyridoxal phosphate in the liver of hypo- and hyperthyroid rats. *Experientia, 23*:719.

Hald, J., Jacobsen, E., and Larsen, V. (1949): Formation of acetaldehyde in the organism in relation to dosage of antabuse (tetraethylthiuramdisulphide) and to alcohol concentration in blood. *Acta Pharmacologica et Toxicologica, 5*: 179.

Hamfelt, A. (1964): Age variation of vitamin B_6 metabolism in man. *Clinica Chimica Acta, 10*:48.

Harris, R. S., Wool, I. G., and Loraine, J. A. (1964): International symposium on vitamin B_6. In: *Vitamin and Hormones,* vol. 22, p. 359. Academic Press, New York.

Himwich, W. A. (1962): Biochemical and neurophysiological development of the brain in the neonatal period. *International Review of Neurobiology, 4*: 117.

Holtz, P., and Palm, D. (1964): Pharmacological aspects of vitamin B_6. *Pharmacological Reviews, 16*: 113.

Huff, J. W., and Perlzweig, W. A. (1944): A product of oxidative metabolism of pyridoxine, 2-methyl-3-hydroxy-4-carboxy-5-hydroxy-methylpyridine (4-pyridoxic acid). *Journal of Biological Chemistry, 155*: 345.

Hurwitz, J. (1952): The enzymic phosphorylation of vitamin B_6 and derivatives and their effects on tyrosine decarboxylase. *Biochemica et Biophysica Acta, 9*:496.

Hurwitz, J. (1953): The enzymatic phosphorylation of pyridoxal. *Journal of Biological Chemistry, 205*: 935.

Hurwitz, J. (1955): Inhibition studies on aldehyde oxidase. *Journal of Biological Chemistry, 212*:757.

Hydén, H., and Lange, P. W. (1970): Do specific biochemical correlates to learning processes exist in brain cells? *Biochemistry of Simple Neuronal Models,* edited by E. Costa and P. Giacobini, p. 317. Raven Press, New York.

Jacobsen, J. G., and Smith, L. H. (1968): Biochemistry and physiology of taurine and taurine derivatives. *Physiological Review, 48*: 424.

Johnson, B. C., Hamilton, T. S., and Mitchell, H. H. (1945): The excretion of pyridoxine, "pseudo-pyridoxine," and 4-pyridoxic acid in the urine and sweat of normal individuals. *Journal of Biological Chemistry, 158*: 619.

Johnson, T. C., and Luttges, M. W. (1966): The effects of maturation on *in vitro* protein synthesis by mouse brain cells. *Journal of Neurochemistry, 13*: 545.

Kahlson, G., and Rosengran, E. (1965): Histamine. *Annual Review of Pharmacology, 5*:305.

Kaplow, L. S. (1955): A histochemical procedure for localizing and evaluating leukocyte alkaline phosphatase activity in smears of blood and marrow. *Blood, 10*: 1023.

Karki, N., Kuntzman, R., and Brodie, B. B. (1962): Storage, synthesis and metabolism of monoamines in the developing brain. *Journal of Neurochemistry, 9*: 53.

Leitch, I., and Hepburn, A. (1961): Pyridoxine: metabolism and requirement. *Nutrition Abstracts and Reviews, 31*: 389.

Lepkovsky, S., Roboz, E., and Haagen-Smit, A. J. (1943): Xanthurenic acid and its role in the tryptophan metabolism of pyridoxine-deficient rats. *Journal of Biological Chemistry, 149*: 195.

Linkswiler, H., and Reynolds, M. A. (1950): Urinary and fecal elimination of B_6 and 4-pyridoxic acid on three levels of intake. *Journal of Nutrition, 41*: 523.

Lolley, R. N., Balfour, W. M., and Samson, F. E., Jr. (1961): The high energy phosphates in developing brain. *Journal of Neurochemistry, 7*: 289.

Loo, Y. H., and Mack K. (1971): Subcellular distribution of B_6 vitamers in cerebral cortex. *Journal of Neurochemistry, 18*: 499–502.

Lovenberg, W., Weissbach, H., and Udenfriend, S. (1962): Aromatic L-amino acid decarboxylase. *Journal of Biological Chemistry, 237*: 89.

Lyon, J. B., Jr., Bain, J. A., and Williams, H. L. (1962): The distribution of vitamin B_6 in the tissues of two inbred strains of mice fed complete and vitamin B_6-deficient rations. *Journal of Biological Chemistry, 237*: 1989.

McClain, L. D., and Bridgers, W. F. (1970): Folate levels in developing mouse brain. *Journal of Neurochemistry, 17*: 763.

McCormick, D. B., Gregory, M. E., and Snell, E. E. (1961): Pyridoxal phosphokinases. I. Assay, distribution, purification, and properties. *Journal of Biological Chemistry, 236*: 2076.

Mandel, P., Bieth, R., and Weill, J. D. (1955): [Research on the energy metabolism of the brain during development in the rat. I. Oxygen consumption II. Acid-soluble phosphorus compounds.] *Bulletin De La Société De Chimie Biologique, 37*:475.

Meister, A. (1965): *Biochemistry of the Amino Acids*, vol. 1, p. 331. Academic Press, New York.

Millichap, J. G. (1957): Development of seizure patterns in newborn animals; significance of brain carbonic anhydrase. *Proceedings of the Society for Experimental Biology and Medicine, 96*:125.

Morisue, T., Morino, Y., Sakamoto, Y., and Ichihara, K. (1960): Enzymic studies on pyridoxine metabolism. III. Pyridoxine phosphate oxidase. *Journal of Biochemistry, 48*:28.

Neff, N. H., Tozer, T. N., and Brodie, B. B. (1967): Application of steady-state kinetics to studies of the transfer of 5-hydroxyindoleacetic acid from brain to plasma. *The Journal of Pharmacology and Experimental Therapeutics, 158*:214.

Norimitsu, K. (1959): The interrelationship between xanthurenic acid formation and amino acid metabolism in the rat. *Kumamoto Medical Journal, 12*:39.

O'Kell, R. T., and Axon, L. L. (1965): Leukocyte alkaline phosphatase in the newborn infant. *American Journal of Obstetrics and Gynecology, 93*:1181.

O'Kell, R.T. (1968): Leukocyte alkaline phosphatase in the infant. *Annals of the New York Academy of Sciences, 155*:980.

Podgell, B. M. (1958): Enzymatic oxidation of pyridoxamine phosphate to pyridoxal phosphate in rabbit liver. *Journal of Biological Chemistry, 232*:761.

Porter, C. C., Clark, I., and Sibler, R. H. (1948): The effect of B vitamin deficiencies on tryptophan metabolism in the rat. *Archives of Biochemistry and Biophysics, 18*:339.

Price, J. M., and Dodge, L. W. (1956): Occurrence of 8-methyl ether of xanthurenic acid in normal human urine. *Journal of Biological Chemistry, 223*:699.

Pritchard, J. A. (1957): Leukocyte phosphatase activity in pregnancy. *Journal of Laboratory and Clinical Medicine, 50*:432.

Rabinowitz, J. C., and Snell, E. E. (1949): Vitamin B_6 group. XV. Urinary excretion of pyridoxal, pyridoxamine, pyridoxine, and 4-pyridoxic acid in human subjects. *Proceedings of the Society for Experimental Biology and Medicine, 70*:235.

Reddy, S. K., Reynolds, M. S., and Price, J. M. (1958): The determination of 4-pyridoxic acid in human urine. *Journal of Biological Chemistry, 233*:691.

Roberts, E., and Frankel, S. (1950): γ-Aminobutyric acid in brain: Its formation from glutamic acid. *Journal of Biological Chemistry, 187*:55.

Roby, C. C., Ober, W. B., and Drorbaugh, J. E. (1956): Pregnanediol excretion in the urine of newborn male infants. *Pediatrics, 17*:877.

Samson, F. E., Jr., and Quinn, D. J. (1967): Na^+-K^+ activated ATPase in rat brain development. *Journal of Neurochemistry, 14*:421.

Sandman, R., and Snell, E. E. (1955): Some effects of ω-methyl analogues of vitamin B_6 in rats. *Proceedings of the Society for Experimental Biology and Medicine, 90*:63.

Saraswathi, S., and Bachhawat, B. K. (1963): Phosphatases from brain. I. Purification and properties of pyridoxal phosphate phosphatase. *Journal of Neurochemistry, 10*:127.

Satoh, K., and Price, J. M. (1958): Fluorometric determination of kynurenic acid and xanthurenic acid in human urine. *Journal of Biological Chemistry, 230*:781.

Schain, R. J., Carver, M. J., Copenhaver, J. V., and Underdahl, N. R. (1967): Protein metabolism in the developing brain: influence of birth and gestational age. *Science, 156*:984.

Schepartz, B. (1963): Oxidation of L-amino acids and incorporation into protein in homogenates of brain at two stages of development. *Journal of Neurochemistry, 10*:825.

Schott, H. F., and Clark, W. G. (1952): Dopa decarboxylase inhibition through the interaction of coenzyme and substrate. *Journal of Biological Chemistry, 196*:449.

Schwartz, R., and Kzeldgaard, N. O. (1951): The enzymatic oxidation of pyridoxal by liver aldehyde oxidase. *Biochemical Journal, 43*:333.

Shimizu, H., Kakimoto, Y., and Sano, I. (1965): Changes in concentration of polyamines in the developing mouse brain. *Nature, 207*:1196.

Snell, E. E. (1953): Summary of known metabolic functions of nicotinic acid, riboflavin and vitamin B_6. *Physiological Review, 33*: 509.

Snell, E. E. (1958): Chemical structure in relation to biological activities of vitamin B_6. In: *Vitamins and Hormones*, vol. XVI, edited by R. S. Harris, G. F. Marrian, and K. V. Thimann, p. 77. Academic Press, New York.

Snell, E. E., Fasella, P. M., Braunstein, A. E., and Rossi Fanelli, A. (1963): *Chemical and Biological Aspects of Pyridoxal Catalysis* (International Union of Biochemistry Symposium, 30). The Macmillan Company, New York.

Sourkes, T. L. (1966): Dopa decarboxylase: Substrates, coenzyme, inhibitors. Pharmacological Reviews, *18*: 53.

Sprince, H., Lowy, R. S., Folsome, C. E., Behrmann, J. S., and Raritan, C. B. (1951): Studies on the urinary excretion of "xanthurenic acid" during normal and abnormal pregnancy: A survey of the excretion of "xanthurenic acid" in normal nonpregnant, normal pregnant, pre-eclamptic, and eclamptic women. *American Journal of Obstetrics and Gynecology, 62*: 84.

Swaiman, K. F., Daleiden, J. M., and Wolfe, R. N. (1970): The effect of food deprivation on enzyme activity in developing brain. *Journal of Neurochemistry, 17*: 1387.

Tapia, R., and Awapara (1967): Formation of γ-aminobutyric acid (GABA) in brain of mice treated with glutamic acid-γ-hydrazide and pyridoxal phosphate γ-glutamyl hydrazone. *Proceedings of the Society for Experimental Biology and Medicine, 126*: 218.

Tapia, R., Pasantes, H., and Massieu, G. (1970): Some properties of glutamate decarboxylase and the content of pyridoxal phosphate in brains of three vertebrate species. *Journal of Neurochemistry, 17*: 921.

Tower, D. B. (1956): Neurochemical aspects of pyridoxine metabolism and function. *The American Journal of Clinical Nutrition, 4*: 329.

Turner, J. M. (1961): Pyridoxal phosphate breakdown by an alkaline-phosphate preparation. *Biochemical Journal, 80*: 663.

Turner, J. M., and Happold, F. C. (1961): Pyridoxamine phosphate-oxidase and pyridoxal phosphate-phosphatase activities in escherichia coli. *Biochemical Journal, 78*: 364.

Umbreit, W. W., and Waddell, J. G. (1949): Mode of action of desoxypyridoxine. Proceedings of *the Society for Experimental Biology and Medicine, 70*: 293.

Vernadakis, A., Curry, J. J., Maletta, G. J., Irvine, G., and Timiras, P. S. (1966): Convulsive responses in prenatally irradiated rats. *Experimental Neurology, 16*: 57.

Vulpe, M., Caldwell, D. F., and Rodin, E. (1963): Ontogenesis of megimide seizures in the rat. *Experimental Neurology, 8*: 350.

Wachstein, M., and Gudaitis, A. (1952*a*): Detection of vitamin B_6 deficiency; utilization of an improved method for rapid determination of xanthurenic acid in urine. *American Journal of Clinical Pathology, 22*: 652.

Wachstein, M., and Gudaitis, A. (1952*b*): Disturbance of vitamin B_6 metabolism in pregnancy. *Journal of Laboratory and Clinical Medicine, 40*: 550.

Wachstein, M. (1956): Evidence for abnormal vitamin B_6 metabolism in pregnancy and various disease states. *American Journal of Clinical Nutrition, 4*: 369.

Wachstein, M., and Lobel, S. (1956): The relation between tryptophane metabolism and vitamin B_6 in various diseases as studied by paper chromatography. *American Journal of Clinical Pathology, 26*: 910.

Wachstein, M., Kellner, J. D., and Ortiz, J. M. (1960): Pyridoxal phosphate in plasma and leukocytes of normal and pregnant subjects following B_6 load tests. *Proceedings of the Society for Experimental Biology and Medicine, 103*: 350.

Wada, H., Morisue, T., Nisimura, Y., Moris, Y., Sakamoto, Y., and Ichihara, K. (1959): Enzymic studies on pyridoxine metabolism. *Proceedings of the Japanese Academy, 35*: 299.

Wada, H., Morisue, T., Moris, Y., Sakamoto, Y., and Ichihara, K. (1960): Enzymic studies on vitamin B_6 metabolism. *Loso Kagaku Shinpojium, 14*: 116.

Wada, H., and Snell, E. E. (1961): The enzymatic oxidation of pyridoxine and pyridoxamine phosphate. *Journal of Biological Chemistry, 236*: 2089.

Waelsch, H., and Lajtha, A. (1961): Protein metabolism in the nervous system. *Physiological Reviews, 41*:709.

White, R. S., and Dempsey, W. B. (1970): Purification and properties of vitamin B_6 kinase from *Escherichia coli. Biochemistry, 9*:4057.

Williams, M. A., and Hata, B. (1959): Liver coenzyme A levels in the vitamin B_6-deficient rat. *Archives of Biochemistry and Biophysics, 80*:367.

Wohl, M. G., Levy, H. A., Szutka, A., and Maldia, G. (1960): Pyridoxine deficiency in hyperthyroidism. *Proceedings of the Society for Experimental Biology and Medicine, 105*:523.

Yamada, K., Sawaki, S., and Hayami, S. (1956): Participation of vitamin B_6 in the biosynthesis of coenzyme A. *Journal of Vitaminology, 2*:296.

Yost, F. J., and Harrison, J. H. (1971): Interaction of pyridoxal 5′-phosphate and malate dehydrogenase. *Biochemical and Biophysical Research Communications, 42*:516.

Advances in Biochemical Psychopharmacology, Vol. 4.
Raven Press, New York © 1972.

Appraisal of Current Methods to Estimate the Turnover Rate of Serotonin and Catecholamines in Human Brain

E. Costa

Laboratory of Preclinical Pharmacology, National Institute of Mental Health, St. Elizabeths Hospital, Washington, D.C. 20032

With the upsurge of interest of psychiatrists and neurologists in the clinical applications of L-3,4-dihydroxyphenylalanine (L-DOPA), the question of the role of pyridoxal-5-phosphate in the control of monoamine turnover in human brain must be reappraised. Reports by Vogel *et al.* (1970) have shown that the aromatic L-amino acid decarboxylase (E.C. 4.1.1.26) (Lovenberg *et al.*, 1962) activity of human brain is very low. If these findings were to be taken at face value, then one should become concerned with the possibility that in the human brain, at least in selected circumstances (administration of decarboxylase inhibitors, vitamin B_6 deficiency, etc.), the aromatic L-amino acid decarboxylase may control serotonin (5-HT), norepinephrine (NE), and dopamine (DM) biosynthesis. Aromatic L-amino acid decarboxylase is a stereospecific enzyme which catalyzes the decarboxylation of several aromatic L-amino acids involved in the biosynthesis of brain 5-HT, NE, and DM (Lovenberg *et al.*, 1962). It is widely distributed in the CNS of several animal species and requires pyridoxal-5′-phosphate as a coenzyme (Holtz, Heise, and Luotke, 1938; see also references in Blaschko, 1945, and Clark, 1963). Although the importance of this enzyme in the biogenesis of NE, DM, and 5-HT is well documented, its regulatory action on the turnover of brain monoamines of many animal species seems doubtful (Brodie *et al.*, 1962). The low concentration of 5-hydroxytryptophan (5-HTP) (Carlsson and Lindqvist, 1970) or L-DOPA (Anton and Sayre, 1964) is below the apparent K_m of aromatic L-amino acid decarboxylase in the brain of laboratory animals. Moreover, the activity of aromatic L-amino acid decarboxylase in brain of several animal species exceeds that of tryptophan-5-hydroxylase or tyrosine-3-hydroxylase by several fold (Udenfriend, 1966). When the interaction of DOPA decarboxylase with its coenzyme (pyridoxal-5-phosphate) is studied in detail, one finds that some pyridoxal phosphate is tightly bound to the apoenzyme whereas another portion is dialyzable (for references, see Sourkes, 1966). The relationship between free coenzyme and coenzyme bound to the apoenzyme may in part control the activity of the aromatic L-amino acid decarboxylase. The role of the bound coenzyme can be revealed indirectly by studying various decarboxylase inhibitors *in vitro*. In these studies, the order of adding supplementary coenzyme, substrate, and inhibitor determined the qualitative type of inhibition. Some of the pyridoxal phosphate added *in vitro* to the aromatic L-amino acid decarboxylase preparation is loosely bound to the apoenzyme and may react nonenzymatically with certain decarboxylase inhibitors, e.g., α-methyl-L-dopa, but a part of the coenzyme is tightly bound and it is not affected in this way. Goldberg *et al.* (1960), Porter *et al.* (1961), and Hess *et al.* (1961) have shown that the effects of these α-methyl phenylalanine analogues on tissue catecholamine levels *in vivo* are due much more to their action on the uptake, binding, and release mechanism of the catecholamines than to an inhibition of the enzyme. Now the theory that the monoamine metabolites of ring hydroxylated α-methylphenylalanine derivatives may function as false neurochemical transmitters by replacing catecholamines or 5-HT in nerve terminals has gained wide acceptance (Carlsson, 1964). In experimental animals, inhibitors of aromatic L-amino acid decarboxylase more potent than α-methyldopa can block aromatic L-amino acid decarboxylase almost completely without lowering the catecholamine content (Brodie *et al.* 1962). Indeed our present understanding of the role of aromatic L-amino acid decarboxylase in adult rat brain does not favor a controlling role of pyridoxal phosphate in monoamine metabolism. However, the studies of Vogel *et al.* (1970) and Lloyd and Hornykiewicz (1970) do not suggest that these data should be extrapolated to human brain because, unlike other species, the activity of aromatic L-amino acid decarboxylase of human brain does not exceed to a great extent that of tyrosine-3-hydroxylase measured *in vitro*. Since we have no information on the tissue concentrations of DOPA and 5-HTP in human brain, and since we cannot rule out that the low concentration of the aromatic L-amino acid decarboxylase is related to some technical problem still to be identified, it is appropriate to refrain from making any final conclusion on the physiological role of this enzyme in human brain. Perhaps immunochemical techniques applied to human brain will give us a better understanding of the role of these enzymes in the synthesis of catecholamines and 5-HT by this tissue (Hartman and Udenfriend, 1970).

I. THE CONTROL OF CATECHOLAMINES AND 5-HT TURNOVER RATE IN RAT BRAIN

The belief that tyrosine-3-hydroxylase is rate limiting in the turnover of catecholamine stores of rats and other experimental animals (Udenfriend, 1966) rests on the following circumstantial evidence:

a) The K_m and velocity of this enzyme are the smallest when compared to that of dopamine-β-hydroxylase and that of aromatic L-amino acid decarboxylase (Udenfriend, 1966).

b) The concentrations of tyrosine in various tissues approach the K_m of tyrosine-3-hydroxylase.

c) In the isolated heart of guinea pig perfused with various radioactive precursors of NE (tyrosine, L-DOPA, and DM), the synthesis rate of NE is proportional to the concentration of tyrosine until the concentration of tyrosine approaches the K_m of tyrosine-3-hydroxylase. Such a restricted relationship does not exist when labeled L-DOPA and DM are used as precursors (Udenfriend, 1966).

d) Blockade of tyrosine-3-hydroxylase by competitive inhibitors leads to an exponential decline of NE and DM brain concentrations; the fractional rate constant (k) of this decline compares to that of the efflux of these amines from brain as measured by isotopic methods (Costa and Neff, 1966; Neff *et al.*, 1969). Partial inhibition of the enzyme leads to a less complete and shorter-duration decrease of the NE and DM brain concentrations (Costa and Neff, 1966).

e) When the tissue concentration of a tyrosine-3-hydroxylase inhibitor is several-fold higher than its K_i, the *in vivo* conversion of ^{14}C-L-tyrosine to NE is completely inhibited (Udenfriend, 1966).

However, we must remember that the K_m for O_2 of tyrosine-3-hydroxylase of the adrenal gland enzyme is greater than the pO_2 of several tissues; hence, it is possible to obtain conditions where the concentrations of this substrate are rate limiting (Costa and Neff, 1970).

The circumstantial evidence that can be marshalled to support a control of 5-HT synthesis by tryptophan-5-hydroxylase is rather limited. Not only is the pO_2 of various brain parts (Cater *et al.*, 1961) below the K_m of crude tryptophan hydroxylase preparations tested *in vitro* (Green and Sawyer, 1966), but also a change of the turnover rate of brain 5-HT related to pO_2 in inhaled gas has been reported (Diaz *et al.*, 1968). In rats, the brain concentrations of 5-HT can be increased by injecting high doses of tryptophan, suggesting that the concentration of this substrate rather than the amount of its hydroxylase is rate limiting. This finding is in keeping with other reports showing that crude preparations of tryptophan-5-hydroxylase tested *in vitro* have a K_m for tryptophan greater than the tissue concentrations of this amino acid (Lovenberg *et al.*, 1968). These findings focus our attention toward the regulation of tryptophan levels as a mechanism to control 5-HT turnover. The unbound concentration of tryptophan is very low in plasma (McMenamy *et al.*, 1957). At physiological pH, tryptophan is bound to a nondialyzable plasma substance, probably protein, but it is released from this binding site upon precipitation of the proteins with trichloroacetic acid. Since the plasma-bound tryptophan can be displaced by drugs (McArthur and Dawkins, 1969), it follows that any study of drug effects on the turnover rate of brain 5-HT must be accompanied by determinations of brain levels of free tryptophan. These measurements are essential to exclude the criticism that a given drug changed the turnover rate of brain 5-HT because it had altered the steady-state equilibrium of tryptophan in brain and plasma.

In practice, our present interest in estimating turnover rates of brain 5-HT, NE, and DM is that of using this measurement to assess whether a drug or a behavioral condition selectively affects the function of the neuronal pathways that contain neurons functioning by releasing one of these amines (Costa and Neff, 1970). It follows that, before extrapolating functional interpretations from biochemical measurements, we must exclude that a change of 5-HT turnover measured in brain is due neither to shifts of pO_2 changes nor to changes of brain tryptophan levels.

II. IS THE ANIMAL MODEL TO MEASURE BRAIN DM, 5-HT AND NE TURNOVER RATE APPLICABLE TO MAN?

If the aromatic L-amino acid decarboxylase in human brain is as low as reported by Vogel *et al.* (1970), the possibility should be considered that in man this enzyme may become rate limiting for 5-HT and catecholamine biosynthesis. This possibility specifically applies when the turnover rates of 5-HT, NE, and DM are measured in some dietary deficiencies and in pathological conditions (Lloyd and Hornykiewicz, 1970*a*). An additional factor to be considered in the therapy with inhibitors of aromatic L-amino acid decarboxylase which form Schiff's base with pyridoxal-5-phosphate is the possibility that a lack of pyridoxal-5-phosphate limits the rate of monoamine biosynthesis. The report by Vogel *et al.* (1970) prompted Lloyd and Hornykiewicz (1970*b*) to re-evaluate the DOPA decarboxylase activity of human brain homogenates. They concluded that homogenates of various areas of the normal human brain can decarboxylate L-DOPA if pyridoxal-5-phosphate is added. The data reported by Lloyd and Hornykiewicz (1970*b*) show that human striatum metabolizes about 0.5 nmoles/mg protein/hr, whereas homogenates of cat striatum tested in similar conditions metabolize about 262 nmoles/mg protein/

hr (Kuntzman *et al.*, 1961). Such a difference may not be explained by the density of dopaminergic axons in the striatum of the two species (Carlsson, 1959; Bertler, 1961); perhaps homogenates of human brain contain a specific inhibitor of the decarboxylase. These findings suggest that the control mechanism of brain monoamine turnover may not be the same in animals and man. In fact, any condition that directly or indirectly affects pyridoxal-5-phosphate concentration and its turnover will probably be reflected more promptly in a change of the activity of aromatic L-amino acid decarboxylase in human brain than in the brain of several experimental animals. Moreover, a controlling role of aromatic L-amino acid decarboxylase could be implemented by drugs which affected the concentrations of the coenzyme. When α-methyldopa is given chronically to man, the incidence of pseudoparkinsonism (Duvoisin, 1968) and a precipitation of psychotic depression (Snyder, 1970) were observed. It is possible that formation of Schiff's bases with endogenous pyridoxal-5-phosphate and/or blockade of human brain decarboxylase are the causes of the pseudoparkinsonism for it is known that α-methylamine metabolites of α-methylphenylalanine derivatives do not bind tenaciously to dopaminergic neurons in caudate nucleus (Gessa *et al.*, 1962).

Analogous considerations apply to the therapeutic use of L-DOPA and DL-5-HTP. Brodie *et al.* (1965) reported that rats receiving large doses of 5-HTP show a reduction of brain DM concentrations. This finding was tentatively explained by suggesting that this amino acid could compete with DOPA for aromatic L-amino acid decarboxylase in dopaminergic neurons and cause formation of 5-HT in nerves that store DM. Indeed, competition for the enzyme by these two substrates had been shown by Rosengren (1960). In addition, an inhibition of tyrosine-3-hydroxylase by large doses of 5-HTP may be considered (Johnson *et al.*, 1968). This finding in animals may have some bearing on the clinical complications observed in man during prolonged administration of L-DOPA or 5-HTP. When L-DOPA is administered to parkinsonians for months and in high doses, choreiform movements are often observed (see Yahr *et al.*, this volume). It is possible that accumulation of DM formed from DOPA in sites where normally another amine is present causes such a symptomatology. Conversely, there is some evidence suggesting a defect in the metabolism of NE, DM, or 5-HT in the brain of patients affected by Huntington's chorea (Markham *et al.*, 1963). When people affected by this disease receive 5-HTP, an exacerbation of choreiform movements ensues (Birkmayer and Hornykiewicz, 1962). Choreiform movements fail to occur in parkinsonian patients receiving 5-HTP (Lee *et al.*, 1968).

III. METHODS USED TO MEASURE TURNOVER RATE OF BRAIN 5-HT, DM, AND NE IN MAN

In the last seven years, many publications have discussed the problems and assumptions associated with turnover-rate measurements of brain monoamines in animals (for references, see Costa, 1970). These uncertainties are further increased in man when one attempts to estimate brain monoamine turnover rate. For example, a number of limitations are imposed by the restrictions of sample collections. Any available method to measure turnover rate of monoamines in human brain can resort only to analyses performed in urine and spinal fluid samples. The sampling of spinal fluid imposes some ethical issues and has some inherent technical limitations determined by the low concentrations of metabolites of these monoamines; certainly these difficulties cannot be bypassed by measuring concentrations of total 5-hydroxyindoles in spinal fluid as reported by Ashcroft *et al.* (1966).

A. Estimation of Brain Monoamine Turnover Rate from Analysis of Acidic Monoamine Metabolites in Spinal Fluid

Active transport mechanisms remove acids from cerebral spinal fluid into blood (Pappenheimer *et al.*, 1961; Prockop *et al.*, 1962). Different acids probably compete for the same transfer process which perhaps is similar to that of renal tubules. Probenecid reduces the renal excretion of organic acids including that of 5-hydroxyindoleacetic acid (5-HIAA) (Despopoulos and Weissbach, 1957). Present knowledge of brain serotonergic mechanisms suggests that 5-HIAA is the major metabolite of brain 5-HT, and occurs in brain with about the same distribution of 5-HT (Roos, 1962; Andén *et al.*, 1963). Homovanillic acid (HVA), a brain metabolite of DM, is distributed in brain as is its amine precursor (Bernheimer, 1964). Injection of probenecid increases the brain concentrations of 5-HIAA (Neff *et al.*, 1964) and HVA (Werdinius, 1967) of various species; since the HVA found in cerebrospinal fluid is derived from brain DM (Bartholini *et al.*, 1966) and 5-HIAA is derived from 5-HT, the rate of increase of these acids in this fluid after maximal effective doses of probenecid might be an index of the metabolism of DM. Therefore, the accumulation of 5-HIAA and HVA after probenecid is used clinically as a presumptive index of monoamine turnover (Korf and Von Praag, 1970).

A mathematical analysis of the changes of 5-HIAA concentrations with time following the blockade of its synthesis with a monoamine oxidase (MAO) inhibitor or of its efflux from brain after probenecid was used to measure the turnover of brain 5-HT in rats (Neff *et al.*, 1964; Tozer *et al.*, 1966; Neff *et al.*, 1967; Neff and Tozer, 1968). These measurements were successfully compared with the turnover rates of brain 5-HT calculated from the accumulation of radioactive 5-HT in the brain of rats infused at constant rate with radioactive tryptophan (Neff *et al.*, 1969) or from the change in specific activity of 5-HT after pulse injections of radioactive tryptophan (Neff *et al.*, 1971). This background information has stimulated various clinical investigators (Dencker *et al.*, 1966) to estimate brain 5-HT and DM turnover from measurements of 5-HIAA and HVA concentrations in spinal fluid. These measurements of the steady state of the amine metabolites in lumbar spinal fluid can be extrapolated to estimate turnover rate of the respective parent amines stored in distant regions of the CNS if the following were true:

1. In different brain parts, the ratio [5-HT] / [5-HIAA] is constant.
2. The rate of spinal fluid formation and efflux is similar to that of 5-HIAA metabolism.
3. Probenecid, a competitive blocker of the acid transport, inhibits the active transport of 5-HIAA and HVA from spinal fluid to plasma irrespective of the concentrations of these acids in the spinal fluid.
4. The concentrations of 5-HIAA and HVA in the lumbar spinal fluid reflect changes in the concentrations of acids occurring in the fluid of cerebral ventricles.
5. The emotional status elicited by spinal tap does not affect differently the concentration of these acids in the spinal fluid of healthy individuals and neuropsychiatric patients.
6. The efficiency of the facilitated transport of the acids is independent of their concentration in spinal fluid; i.e., the acid transport proceeds at a maximal velocity independently of the concentration of the substrate.

Many of these assumptions may not be true. The ratio of [5-HT]/[5-HIAA] is not uniform in different brain parts (Ashcroft *et al.*, 1968); the clearance of inulin from spinal fluid is half that of 5-HIAA (Ashcroft *et al.*, 1968); and the competitive inhibition of efflux of acids from spinal fluid elicited by probenecid depends on the initial concentrations of acid in spinal fluid (Bowers, 1969). Furthermore, in dogs the effect of probenecid on the concentrations of HVA and 5-HIAA in spinal fluid is different in different parts of the CNS (Goldberg *et al.*, 1966).

Despite these and other discrepancies, important changes in the turnover rate of brain monoamines appear to be related to the concentrations of the respective acids in lumbar spinal fluid. The administration of α-methyltyrosine, which inhibits the synthesis of catecholamines, reduces the concentrations of HVA in the spinal fluid of the cat (Bowers, 1970). Patients with either parkinsonism or presenile dementia evince a significant decrease in the concentration of 5-HIAA and HVA in spinal fluid along with a concomitant reduction of brain DM (Gottfries *et al.*, 1969*a*, *b*). Weiner *et al.* (1969) reported that administration of L-DOPA to parkinsonian patients is associated with an increase of the concentrations of HVA in the spinal fluid. In view of this evidence, reports that 5-HIAA concentrations in spinal fluid of patients with affective disorders (Dencker *et al.*, 1966) are reduced should be taken as an indication of possible changes of brain 5-HT turnover. However, because of the assumptions listed above, a negative finding does not exclude the possibility that monoamine turnover rate is altered. Nevertheless, the analysis of the available data allows us to make some constructive suggestions regarding the future course of action. Korf and Von Praag (1970) have suggested that intravenous administration of probenecid (1 g in 250 ml during 20 min) can eliminate the disadvantages associated with the variation in distribution of the drug reported after oral administration. Granting that this new approach eliminates some of the problems, the basic difficulty of measuring the efflux of 5-HIAA from various parts of the CNS from single-point analysis of the 5-HIAA content of lumbar spinal fluid still remains unsolved. In view of this difficulty, it may appear more constructive to resort to other methods which limit the number of necessary assumptions. These methods should be preferred if they yield equal or better information without raising the ethical issues associated with repeated spinal punctures.

B. Estimation of Brain Monoamine Turnover Rate from Urinary Analysis of their Metabolites

A prolonged elevation in the urinary excretion of catecholamines, 5-hydroxyindolealkylamines, and their metabolites is the best measurement of an increased synthesis of these neurotransmitters. However, a study of the changes of the urinary concentrations of catecholamines or 5-hydroxytryptamine and their metabolites does not yield information on the turnover rates of monoamines in individual organs such as heart, brain, and adrenal gland. To answer the question if the turnover rates of brain 5-HT, NE, and DM change either during disturbances of affect and emotion (Schildkraut and Kety, 1967) or in various neuropathological cases, it is important to distinguish if the monoamine output from brain is preferentially enhanced. Bigelow *et al.* (1969) demonstrated that an increase of the turnover rate of catecholamines in an organ is accompanied by an increase in the rate of urinary excretion of NE and its metabolites. They calculated that in a rat weighing 200 g the steady-state concentration of the body catecholamines is about 10μg. However, the daily urinary excretion is about 13 times greater than the pool size of the body NE content. These estimates suggest an average turnover time of endogenous NE equal to about 110 min. Personal communications with Spector and Bigelow suggest that the NE in blood vessels could account for about 76% of the daily urinary output of catecholamines. This output for the greatest part consists of glycols which in the normal rat account for over 80% of the total urinary output of catechols. From the data presented by Bigelow *et al.* (1969), one might extrapolate that the brain contribution to this urinary catechol excretion accounts for about 20% of the total excretion. In man, the glycols account for roughly 30% of the daily catecholamine output, and, therefore, these figures must be modified.

Can we isolate and identify the contribution by brain to the urinary daily output of catechols in man? At this moment, we can merely discuss working hypotheses. Our laboratory is presently entertaining a hypothesis based on the following assumptions: (1) the peripheral contribution to the formation of the urinary pools of NE, DM, and 3-methoxy-4-hydroxyphenylglycol (MHPG) can be identified by perturbing sympathetic function with bethanidine; (2) bethanidine changes the urinary excretion of catechols because it reduces sympathetic function and inhibits intraneuronal MAO; and (3) the entity of the perturbation determined by bethanidine can be assessed by injecting ^{3}H-NE and measuring the change in urinary radioactive catechols caused by the perturbation. To carry on the measurements indicated in (3), the labeling of peripheral NE stores with ^{3}H-NE seems in order. We propose that, beginning 16 hr after labelling, 24-hr urine specimens are analyzed for various radioactive and nonradioactive catechol compounds. One week later the labeling is repeated, but this time bethanidine (10 mg/kg twice daily) is injected immediately after labeling. Collections of 24-hr urine specimens for catechol assay are also performed. In both experiments, the specific radioactivities of epinephrine, NE, metanephrine, normetanephrine, and MHPG are assayed. The estimation of brain NE turnover rate (NET_R) is performed as shown in equation (1):

$$NET_R = NET - NEP_0 \tag{1}$$

NET_R = μmole of MHPG turnover by the brain in 24 hr

and

$NE\ P_0$ = μmole of MHPG turned over by peripheral tissue in 24 hr

and

NET = total urinary output of MHPG during 24 hr

then

ΔA_0 = μmole/24 hr of MHPG before bethanidine − μmole/24 hr of MHPG after bethanidine

$$NEP_0 = \Delta A_0 \frac{\text{MHPG (dpm/24 hr) before bethanidine}}{\text{MHPG (dpm/24 hr) after bethanidine}}$$

We are presently experimenting with rats where direct analyses of the turnover rate of brain NE can be performed simultaneously with urinary analysis. With this method we are attempting to evaluate the discrepancy between direct *in vivo* measurement of brain NE turnover and the turnover rate of brain NE calculated from the urinary output by the perturbation method described above.

REFERENCES

Andén, N. E., Roos, B. E., and Werdinius, B. (1963): On the occurrence of homovanillic acid in brain and cerebrospinal fluid and its determination by a fluorometric method. *Life Sciences, 2*:448.

Anton, A. H., and Sayre, D. F. (1964): The distribution of dopamine and Dopa in various animals and a method for their determination in diverse biological material. *Journal of Pharmacology and Experimental Therapeutics, 145*:326.

Ashcroft, G. W., Crawford, T. B. B., Eccleston, D., Sharman, D. F., MacDougall, E. J., Stanton, J. B., Binns, J. K. (1966): 5-Hydroxyindole compounds in the cerebrospinal fluid of patients with psychiatric or neurological diseases. *The Lancet, 2*:1049.

Ashcroft, G. W., Moir, A. T. B., and Eccleston, D. (1968): Discussion of *in vivo* measurement of brain serotonin turnover. *Advances in Pharmacology, 6A*:110.

Bartholini, G., Pletscher, A., and Tissot, R. (1966): On the origin of homovanillic acid in the cerebrospinal fluid. *Experientia, 22*:609.

Bernheimer, H. (1964): Distribution of homovanillic acid in the human brain. *Nature, 204*:587.

Bertler, A. (1961): Occurrence and localization of catecholamines in the human brain. *Acta Physiologica Scandinavica, 51*:97.

Bigelow, L. B., Dairman, W., Weil-Malherbe, H., and Udenfriend, S. (1969): Increased synthesis of catecholamines and their metabolites following the administration of phenoxybenzamine. *Molecular Pharmacology, 5*:565.

Birkmayer, W., and Hornykiewicz, O. (1962): Der L-dioxyphenylalanine (L-dopa) effect beim Parkinson syndrom des menschen: zur pathogenese und behandlung Parkinson-Akinese. *Arch. Psychiat. New., 203*:560.

Blaschko, H. (1945): The amino acid decarboxylases of mammalian tissue. *Advances in Enzymology, 5*:67.

Bowers, M. B., Jr. (1969): Deficient transport mechanism for the removal of acid monoamine metabolites from cerebrospinal fluid. *Brain Research, 15*:522.

Bowers, M. B., Jr. (1970): CSF homovanillic acid: Effect of probenecid andα-methyltyrosine. *Life Sciences, 9*:691.

Brodie, B. B., Kuntzman, R., Hirsch, C. W., Costa, E. (1962): Effects of decarboxylase inhibition on the biosynthesis of brain monoamines. *Life Sciences, 1*:81.

Brodie, B. B., Comer, M. S., Costa, E., and Dlabac, A. (1965): The role of brain serotonin in the mechanism of the central action of reserpine. *Journal of Pharmacology and Experimental Therapeutics, 152*:340.

Carlsson, A. (1959): The occurrence, distribution and physiological role of catecholamines in the nervous system. *Pharmacological Reviews, 11*:490.

Carlsson, A. (1964): Functional significance of drug-induced changes in brain monoamine levels. In: *Progress in Brain Research of Biogenic Amines*, Vol. 8, edited by H. E. Himwich and W. A. Himwich, pp. 9–27. Elsevier, Amsterdam.

Carlsson, A., and Lindqvist, M. (1970): Accumulation of 5-hydroxytryptophan in mouse brain after decarboxylase inhibition. *Journal of Pharmacy and Pharmacology, 22*:726.

Cater, D. B., Garattini, S., Marina, F., and Silver, I. A. (1961): Changes of oxygen tension in brain and somatic tissues induced by vasodilator and vasoconstrictor drugs. *Proceedings of the Royal Society of London, Series B, 155*:136.

Clark, W. G. (1963): Inhibition of aminoacid decarboxylase. In: *Metabolic Inhibitors,* Vol. 1, edited by R. M. Hochster and J. H. Quastel, pp. 315–381. Academic Press, New York.

Costa, E. (1970): Simple neuronal models to estimate turnover rate of noradrenergic transmitters *in vivo.* In: *Biochemistry of Simple Neuronal Models; Advances in Biochemical Psychopharmacology,* Vol. 2, edited by E. Costa and E. Giacobini, pp. 169–204. Raven Press, New York.

Costa, E., and Neff, N. H. (1966): Isotopic and nonisotopic measurements of the rate of catecholamine biosynthesis. In: *Biochemistry and Pharmacology of the Basal Ganglia,* edited by E. Costa, L. Cote, and M. D. Yahr, pp. 141–156. Raven Press, New York.

Costa, E., and Neff, N. H.: (1970): Estimation of turnover rates to study the metabolic regulation of the steady-state level of neuronal monoamines. In: *Handbook of Neurochemistry,* Vol. 4, edited by A. Lajtha, p. 45. Plenum Publ. Co., New York.

Dencker, S. J., Malm, U., Roos, B. E., and Werdinius, B. (1966): Acid monoamine metabolites of cerebrospinal fluid in mental depression and mania. *Journal of Neurochemistry, 13*:1545.

Despopoulos, A., and Weissbach, H. (1957): Renal metabolism of 5-hydroxy-indoleacetic acid. *American Journal of Physiology, 189*:548.

Diaz, P. M., Ngai, S. H., and Costa, E. (1968): Effect of oxygen on brain serotonin metabolism in rats. *American Journal of Physiology, 214*:591.

Duvoisin, R. C. (1968): Neurological reactions to psychotropic drugs. In: *Psychopharmacology: A Review of Progress 1957–1967,* edited by D. H. Efron, pp. 561–573. Superintendent of Documents, U. S. Government Printing Office, Washington, D. C.

Gessa, G. L., Costa, E., Kuntzman, R., and Brodie, B. B. (1962): On the mechanism of norepinephrine release by α-methylmetatyrosine. *Life Sciences, 1*:353.

Goldberg, L. I., DaCosta, F. M., and Ozaki, M. (1960): Actions of the decarboxylase inhibitor, α-methyl-3-4-dihydroxyphenylalanine, in the dog. *Nature, 188*:502.

Gottfries, C. G., Gottfries, I., and Roos, B. E. (1969a): The investigation of homovanillic acid in the human brain and its correlation to senile dementia. *The British Journal of Psychiatry, 115*:563.

Gottfries, C. G., Gottfries, I., and Roos, B. E. (1969*b*): Homovanillic acid and 5-hydroxyindoleacetic acid in the cerebrospinal fluid of patients with senile dementia, presenile dementia and parkinsonism. *Journal of Neurochemistry, 16*:1341.

Green, H., and Sawyer, J. L. (1966): Demonstration, characterization and assay procedure of tryptophan hydroxylase in rat brain. *Analytical Biochemistry, 15*:53.

Guldberg, H. C., Ashcroft, G. W., and Crawford, T. B. B. (1966): Concentrations of 5-hydroxyindoleacetic acid and homovanillic acid in the cerebrospinal fluid of the dog before and during treatment with probenecid. *Life Sciences, 5*:1571.

Hartman, B. K., and Udenfriend, S. (1970): Immunofluorescent localization of dopamine β-hydroxylase in tissues. *Molecular Pharmacology, 6*:85.

Hess, S. M., Connamacher, R. H., Ozaki, M., and Udenfriend, S. (1961): The effects of α-methyldopa and α-methyl-meta-tyrosine on the metabolism of norepinephrine and serotonin *in vivo. Journal of Pharmacology and Experimental Therapeutics, 134*:129.

Holtz, P., Heise, R., and Luotke, K.: (1938): Fementativer Abbau Von L-Dioxyphenylalanine (DOPA) Durch Niere. *Archives Experimentelle Patologie und Pharmakologie, 191*:87.

Johnson, G. A., Kim, E. G., and Boukma, S. J. (1968): Mechanism of norepinephrine depletion by 5-hydroxytryptophan. *Proceedings of the Society of Experimental Biology and Medicine, 128*:509.

Korf, J., and Von Praag, H. M. (1970): The intravenous probanecid test: A possible aid in evaluation of the serotonin hypothesis on the pathogenesis of depression. *Psychopharmacologia, 18*:129.

Kuntzman, R., Shore, P. A., Bogdanski, D., and Brodie, B. B. (1961): Microanalytical procedures for fluorometric assay of brain dopa, 5-HTP decarboxylase, norepinephrine and serotonin and a detailed mapping of decarboxylase activity in brain. *Journal of Neurochemistry, 6*:226.

Lee, D. K., Markham, C. H., and Clark, W. G. (1968): Serotonin (5-HT) metabolism in Huntington's Chorea. *Life Sciences, 7*:707.

Lloyd, K., and Hornykiewicz, O. (1970*a*): Parkinson's disease: activity of L-DOPA decarboxylase in discrete brain regions. *Science, 170*:1212.

Lloyd, K., and Hornykiewicz, O. (1970*b*): Occurrence and distribution of L-DOPA decarboxylase in the human brain. *Brain Research, 22*:426.

Lovenberg, W., Jequier, E., and Sjoerdsma, A. (1968): Tryptophan hydroxylation in mammalian systems. In: *Advances in Pharmacology,* Vol. 6A, edited by S. Garattini, P. A. Shore, E. Costa, and M. Sandler, pp. 21–36. Academic Press, New York.

Markham, C. H., Clark, W. G., and Winters, W. D. (1963): Effect of alphamethyl dopa and reserpine in Huntington's Chorea, Parkinson's Disease and other movement disorders. *Life Sciences, 2*:697.

McArthur, J. N., and Dawkins, P. D. (1969): The effect of sodium salicylate on the binding of L-tryptophan to serum proteins. *Journal of Pharmacy and Pharmacology, 21*:744.

McMenamy, R. H., Lund, C. C., and Oncley, J. L. (1957): Unbound amino acid concentrations in human blood plasma. *Journal of Clinical Investigations, 36*:1672.

Neff, N. H., Lin, R. C., Ngai, S. H., and Costa, E. (1969): Turnover rate measurements of brain serotonin in unanesthetized rats. In: *Advances in Biochemical Psychopharmacology,* Vol. 1, edited by E. Costa and P. Greengard, pp. 91–109. Raven Press, New York.

Neff, N. H., Ngai, S. H., Wang C., and Costa, E. (1969): Calculation of the rate of catecholamine synthesis from the rate of conversion of ^{14}C tyrosine to catecholamines: Effect of adrenal demedullation on synthesis rates. *Molecular Pharmacology, 5*:90.

Neff, N. H., Spano, P. F., Groppetti, A., Wang, C. T., and Costa, E. (1971): A simple procedure for calculating the synthesis rate of norepinephrine, dopamine and serotonin in rat brain. *Journal of Pharmacology and Experimental Therapeutics, 176*:701.

Neff, N. H., Tozer, T. N., and Brodie, B. B. (1964): Indole metabolism, Part II. A Specialized Transport System to Transfer 5-HIAA directly from Brain to Blood. *The Pharmacologist, 6*:194.

Neff, N. H., Tozer, T. N., and Brodie, B. B. (1967): Application of steady-state kinetics to studies of the transfer of 5-HIAA from brain to plasma. *Journal of Pharmacology and Experimental Therapeutics, 158*:214.

Neff, N. H., and Tozer, T. N. (1968): In vivo measurement of brain serotonin turnover. *Advances in Pharmacology, 6A*:97.

Pappenheimer, J. R., Heisey, S. R., and Jordan, E. F. (1961): Active transport of Diodrast and phenolsulphonphtalein from cerebrospinal fluid to blood. *American Journal of Physiology, 200*:1.

Porter, C. C., Totaro, J. A., and Leiby, C. M. (1961): Some biochemical effects of α-methyl-3-4-dihydroxyphenylalanine and related compounds in mice. *Journal of Pharmacology and Experimental Therapeutics, 134*:139.

Prockop, L. D., Schanker, L. S., and Brodie, B. B. (1962): Passage of lipid-insoluble substances from cerebrospinal fluid to blood. *Journal of Pharmacology and Experimental Therapeutics, 135*:266.

Roos, B. E. (1962): On the occurrence and distribution of 5-hydroxyindole-acetic acid in brain. *Life Sciences, 1*:25.

Rosengren, E. (1960): Are dihydroxyphenylalanine decarboxylase and 5-hydroxytryptophan decarboxylase individual enzymes? *Acta Physiologica Scandinavica, 49*:364.

Schildkraut, J. J., and Kety, S. S. (1967): Biogenic amines and emotion. *Science, 156*:21.

Snyder, S. (1970): Catecholamines, brain function and how psychotropic drugs act. In: *Principles of Psychopharmacology,* edited by W. G. Clark and J. del Gindice, pp. 115–124. Academic Press, New York.

Sourkes, T. L. (1966): Dopa decarboxylase: Substrates, coenzyme, inhibitors. *Pharmacological Reviews, 18*:53.

Tozer, T. N., Neff, N. H., and Brodie, B. B. (1966): Application of steady-state kinetics to determine turnover rate and turnover time of brain serotonin in normal and reserpine treated rats. *Journal of Pharmacology and Experimental Therapeutics, 153*:177.

Udenfriend, S. (1966): Tyrosine hydroxylase. *Pharmacological Reviews, 18*:43.

Vogel, W. H., McFarland, H., and Prince, L. N. (1970): Decarboxylation of 3,4-dihydroxyphenylalanine in various human adult and fetal tissues. *Biochemical Pharmacology, 19*:618.

Weiner, W., Harrison, W., and Klawans, H. (1969): L-dopa and cerebrospinal fluid homovanillic acid in parkinsonism. *Life Sciences, 8*:971.

Werdinius, B. (1967): Effect of probenecid on the levels of monoamine metabolites in the rat brain. *Acta Pharmacologica et toxicologica, 25*:18.

Advances in Biochemical Psychopharmacology, Vol. 4.
Raven Press, New York © 1972.

Pyridoxine, DOPA, and Parkinsonism

Melvin D. Yahr, Roger C. Duvoisin, Lucien Côté, and Gerald Cohen

Clinical Center for Research in Parkinson's and Allied Diseases, and the Department of Neurology, College of Physicians and Surgeons, Columbia University, New York, New York 10032.

The use of pyridoxine hydrochloride in the treatment of Parkinson's disease dates back some 30 years, soon after it became available as a therapeutic agent. Prompted by the theoretical suggestion that vitamin B_6 might play a role in muscle metabolism (Antopol and Schotland, 1940), by the pathological findings of spinal cord degeneration in animals fed a diet deficient in pyridoxine (Zimmerman, 1943), and by the clinical observation of increased muscle strength in pellagarans given pyridoxine (Spies, Bean, and Ashe, 1939), a number of clinical trials were attempted. Administered by a variety of routes —orally, intravenously, and intrathecally in doses of 50–100 mg, either on a daily or weekly basis—pyridoxine underwent extensive clinical testing. Initial claims of a favorable effect (Jollife, 1940; Baker, 1941) were refuted by subsequent negative reports (Barker, Stein, Miller, and Wintrobe, 1941; Zeligs, 1941), and pyridoxine was abandoned as a treatment for parkinsonism. Evidence that dopamine which is normally present in basal ganglia structures, particularly the striatum, is depleted in parkinsonism (Ehringer and Hornykiewicz, 1960) renewed interest in pyridoxine in this disorder. Since the decarboxylation of dihydroxyphenylalanine (DOPA) to form dopamine is catalyzed by a pyridoxal-dependent enzyme, L-aromatic amino acid decarboxylase, a deficiency in this enzyme had been suggested as the possible underlying cause of the striatal dopamine loss in parkinsonism. Recently, evidence for a reduction of decarboxylase activity in Parkinson's disease has been presented by Metzel, Weinmann, and Reichert (1969). Using biopsy material obtained from the thalamus and pallidum of parkinson patients, they showed that its decarboxylase activity was low, and that, when incubated with pyridoxal phosphate, it increased by 50%. From this, it was argued that pyridoxine might enhance the formation of dopamine in brain. Although some investigators (Finke, 1954; Hartmann-von Monakow, 1961; Sigwald, 1957) have reported beneficial effects on the symptons of parkinsonism following administration of large oral doses of pyridoxine (800–1200 mg daily), the results appear far from conclusive. In a double-blind study, one of us (R.C.D.) found no significant difference between those receiving 1000 mg daily of pyridoxine hydrochloride and those receiving a matching placebo. Hence, we concluded that the administration of pyridoxine was ineffective in relieving the symptoms of parkinsonism.

The introduction of L-DOPA as a therapeutic agent for parkinsonism (Cotzias, Papavasiliou, and Gellene, 1969; Yahr, Duvoisin, Schear, Barrett, and Hoehn, 1969) has again focused attention on pyridoxine. The central role of pyridoxal phosphate in its metabolism has led to a number of important yet opposing viewpoints regarding the long-term administration of L-DOPA. On one hand, it has been suggested that the effect of exogenously administered DOPA could be enhanced by the addition of pyridoxine; on the other hand, apprehension has been expressed that its continued usage might induce a deficiency state of this essential vitamin. The present study was undertaken to seek answers to both of these problems.

For the purpose of this study, patients were selected from an ongoing investigation of the therapeutic efficacy of L-DOPA in parkinsonism begun some three years ago. Two groups of patients were studied: (a) those to whom pyridoxine was added to an existing regimen of L-DOPA and assessment of its effects on their parkinson symptoms and the metabolic fate of L-DOPA determined, and (b) patients receiving L-DOPA in whom all supplemental vitamin therapy was withheld and clinical evidence of vitamin B_6 deficiency sought and measurements relative to blood levels of vitamin B_6 determined.

I. COMBINED USE OF L-DOPA AND PYRIDOXINE

Thirty patients, all with Parkinson's disease, who were receiving L-DOPA in maximally tolerable dosage for extended intervals of time, were given pyridoxine orally in doses of 5–1000 mg a day. In special instances, 100–200 mg of pyridoxine was given intramuscularly. In eight patients, plasma levels of DOPA and homovanillic acid (HVA) were determined before and during pyridoxine administration. In these patients, dietary control was observed; fruits, vegetables, nuts, and aspirin-containing drugs were withheld during the preceding 12 hr and during the period of testing. All patients received their regular dosage of L-DOPA which varied from 3-1/2 to 6 g/day, and during a 1–3 day period they received pyridoxine in daily dosage varying from 100–1000 mg. Blood samples were drawn before the morning dose of L-DOPA and pyridoxine, and then hourly thereafter for a period of 7 hr. These samples were collected in dry heparinized tubes and immediately centrifuged; the plasma was deproteinized with perchloric acid in a final concentration of 0.4 N, chilled for 15 min, and centrifuged.

A portion of the clear supernatant was assayed for HVA by the method of Andén (Andén, Roos, and Werdinius, 1963). From a separate portion of clear supernatant, plasma DOPA and other catechols were extracted on alumina by the method of Anton and Sayre (1962) and then DOPA separated itself on Dowex-50 (hydrogen-form) as described by Bertler, Carlsson, and Rosengren (1958). DOPA was then converted to a fluorophore by the method of Chang (1964) and measured spectrofluorometrically. Plasma samples to which known amounts of DOPA and HVA were added were processed with each set of assays. The values for the patients were calculated from these internal standards.

In every patient given pyridoxine, a reduction of the therapeutic response to L-DOPA was noted. In no patient was the response to L-DOPA enhanced. The effect of larger doses of pyridoxine was more rapid, greater, and sustained for a longer period of time after discontinuation than those of lower doses. In seven patients who were given 800–1000 mg daily of pyridoxine, a loss of beneficial effects of L-DOPA were apparent within one day. By the second day one could hardly notice any effect of L-DOPA, and by the third day they had reverted to a parkinson state similar to that prior to receiving L-DOPA. Following discontinuation of pyridoxine, 7 to 10 days were required for these patients to fully regain their response to L-DOPA.

One of the major limiting side effects produced by L-DOPA is abnormal involuntary movements (Yahr, 1970). These movements which may be of a choreatic, athetotic, or dystonic nature reduce the functional capacity of patients and restrict the usefulness of L-DOPA. It had been suggested by Jameson (1970) that these effects could be relieved by pyridoxine without abolishing the benefits of L-DOPA. Six patients in the above group had involuntary movements which were markedly reduced or abolished by pyridoxine but only at the expense of a diminished overall benefit in their parkinson state. The administration of 100–400 mg of pyridoxine per day to six patients gave similar results. In 10 patients in whom 50–75 mg per day of pyridoxine was administered and in whom abnormal involuntary movements were predominant, the following sequence occurred. At first there was marked disappearance of these movements while the beneficial effects of L-DOPA appeared to be maintained. However, within several days a total loss of the beneficial effect of L-DOPA became apparent. In a further attempt to suppress the abnormal movements but maintain the beneficial effects, still smaller doses of pyridoxine hydrochloride ranging from 5 to 30 mg were administered to 21 patients. Though the effects were milder and more insidious in onset, most patients lost their response to L-DOPA necessitating the withdrawal of pyridoxine within a week. In only six instances, patients continued to use pyridoxine on an intermittent basis for periods up to six months feeling that it reduced the intensity of their abnormal involuntary movements. In each of these, there was some return of their parkinsonism and essentially pyridoxine was virtually abandoned. However, an occasional patient resorted to its use, on an intermittent basis, when the need to keep movements at a minimal level existed, knowing fully well that some return of parkinsonian symptomatology would occur.

In eight of the patients in the preceding group, plasma levels of DOPA and HVA were determined prior to and following administration of pyridoxine hydrochloride. In patients receiving DOPA, it was not usually detected in morning blood specimens which were drawn 11 to 12 hr after the dose of L-DOPA given the previous evening. Following the oral administration of 1 g of L-DOPA, measurable plasma concentrations of DOPA were found which attained peak values usually within 1 to 2 hr and declined rapidly during the third and fourth hours. Following the administration of pyridoxine, however, the same dose of L-DOPA resulted in much lower levels of plasma DOPA. The effect of pyridoxine on the plasma level of DOPA is illustrated in Fig. 1 showing the hourly levels of DOPA following an oral dose of L-DOPA with and without the concomitant administration of pyridoxine.

The data for all patients studied are summarized in Table 1. A similar effect was found in all cases studied although the levels of DOPA varied consid-

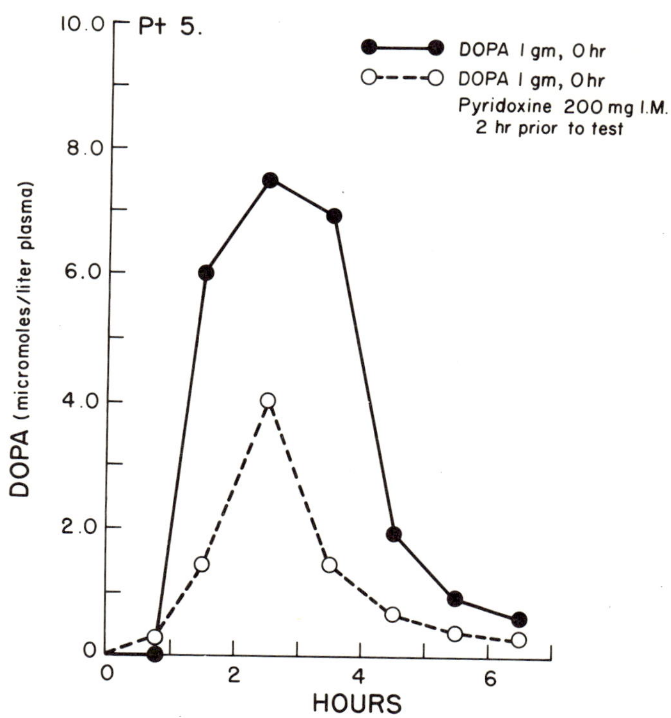

FIG. 1. Hourly plasma DOPA levels in a representative case (patient H. W.) following 1 g of L-DOPA (solid line) and following 1 g of L-DOPA plus pyridoxine (dotted line). This patient received 200 mg of pyridoxine HCl by intramuscular injection 2 hr prior to the administration of L-DOPA.

erably. In every case the administration of pyridoxine was associated with lower plasma levels of DOPA. The magnitude of this effect showed no correlation with the amount of pyridoxine given; however, two patients, H. L. and A. K., who received their last dose of pyridoxine the day before the test, showed a less striking depression in the plasma DOPA level. Indeed, the differences in the peak plasma DOPA levels attained with and without pyridoxine treatment in patient A. K. are probably not significant.

Plasma HVA levels rose after the administration of L-DOPA and remained elevated throughout the entire period of observation. The mean of the hourly values for each patient is presented in Table 1. Plasma HVA did not seem to be appreciably altered by pyridoxine in four patients but was significantly elevated in the remainder.

II. PYRIDOXINE DEFICIENCY AND L-DOPA

It is generally agreed that vitamin B_6 is essential for normal metabolic functions although clinical manifestations relative to its deficiency have been incompletely and rarely documented. Lesions of the skin and mucous membranes, anemia, renal calculi, and convulsive seizures have been described in individuals suspected of being pyridoxine-deficient as a result either of dietary restriction or of the use of metabolic antagonists of pyridoxine. During the past three years, we have treated over 500 parkinsonian patients with L-DOPA and have not encountered any physical or laboratory abnormalities indicative of nutritional deficiency. Specifically, no lesion of the type attributable to a lack of vitamin B_6 has been found. However, to determine that a relative deficiency of vitamin B_6 without overt symptoms did not exist, blood levels of pyridoxal phosphate were determined in six patients.

Measurements were performed with a modification of a standard method (Umbreit, Bellamy, and Gunsalus, 1945) in which the activity of a vitamin B_6-deficient preparation of tyrosine decarboxylase was stimulated by the pyridoxal phosphate present in heparinized blood plasma. The evolution of $^{14}CO_2$ from $^{14}C_1$-tyrosine was measured. Fasting blood specimens were drawn in heparinized Vacutainers and immediately immersed in ice water. Samples were centrifuged, and the supernatant plasma was used for analyses of the pyridoxal phosphate. Assays were performed in Erlenmeyer flasks with built-in center wells. The center well contained 0.2 ml of 40% NaOH to trap the CO_2. The main compartment contained 1.0 ml of 0.5 M acetate buffer at pH 5.5, 0.4 ml of a freshly dissolved solution (2.5 mg/ml) of tyrosine decarboxylase apoenzyme (Worthington), and 1.0 ml of heparinized plasma. The mixture was equilibrated at 37°C for 1 hr in order to allow the apoenzyme and the pyridoxal phosphate co-factor to interact. Then 0.5 ml of 0.03 M tyrosine [containing 3/nmoles/ml of $^{14}C_1$-tyrosine (New England Nuclear), sp. act. 33.6mc/mole] was added, and the system was tightly capped with a rubber

TABLE 1. *Effect of pyridoxine (B_6) on plasma levels of DOPA and HVA in parkinson patients on L-DOPA therapy.*

Pt.	g/day L-DOPA*	Amount B_6 Administered (mg)				Peak Values Plasma DOPA (μM)				HVA (μM)	
		prior to test		day of test		L-DOPA ingested at 9:00 AM		L-DOPA ingested at 1:00 PM		mean values	
		2 days	1 day	9:00 AM	1:00 PM	$\bar{s}B_6$	$\bar{c}B_6$	$\bar{s}B_6$	$\bar{c}B_6$	$\bar{s}B_6$	$\bar{c}B_6$
BC	5.0	1000	1000	250	250	18.2	8.9	28.3	8.7	9.4	9.0
TK	5.0	0	1000	250	250	10.3	0.8	6.5	2.9	21.0	17.4
HF	6.0	800	800	200	200	19.1	5.4	17.2	1.1	19.5	19.8
HL	5.5	200	200	0	0	11.1	6.7	4.1	4.6	19.3	37.8
DV	5.0	100	100	25	25	12.6	3.9	28.6	5.7	33.3	40.5
AK	3.5	100	100	0	0	16.8	14.5	11.5	13.0	25.4	25.3
MA	5.0	0	200⁺	200⁺	–	8.1	2.8	–	–	13.9	24.5
HW	5.0	0	0	200⁺	–	7.5	4.0	–	–	15.2	19.3

*The daily dose of L-DOPA and the amount given on the day of the tests were the same prior to and after the administration of Pyridoxine.
⁺Pyridoxine was administered intramuscularly instead of orally, as in the other patients.

stopper. After 30 min, 0.2 ml of 1N H_2SO_4 was added by syringe through the cap to stop the enzymatic reaction. Incubation was continued at 37°C for another hour to permit the $^{14}CO_2$ to diffuse and become entrapped in the NaOH. A 0.1 ml sample of the NaOH was removed and added to 15 ml of Bray's solution and 3 ml of Tris buffer, pH 8.0, in a counting vial. Radioactivity was analyzed on a Packard Tricarb Scintillation Spectrometer. For each sample, an internal standard of 5 ng of pyridoxal phosphate was run. A blank value was obtained by analyzing the enzyme without added plasma. Our range of normal values was similar to those of Hamfelt (1967) who used a similar method, except that we measured the formation of radioactive tyramine rather than $^{14}CO_2$.

In nine blood samples from six hospitalized parkinson patients receiving doses of 1.5–5.0 g L-DOPA per day for extended periods of time, an average blood pyridoxal phosphate level of 5.9 $\pm$ 3.3 (a.d.) ng/ml (range, 2.6–11.5 ng/ml) was observed. Single blood samples from 10 control patients drawn from the same population but hospitalized for a variety of reasons and not under treatment with L-DOPA showed an average blood pyridoxal phosphate level of 5.8 $\pm$ 2.8 (a.d.) ng/ml (range, 1.1 – 12.5 ng/ml). These data indicate no effect of L-DOPA therapy on blood pyridoxal phosphate levels. In addition, three patients were studied before and after the initiation of L-DOPA therapy. Two showed no change: 9.8 (before) vs. 10.5 ng/ml (after), and 8.0 (before) vs. 7.5 ng/ml (after). One showed a change from 6.4 (before) to 4.0 ng/ml (after); however, the plasma level was still in the normal range. These data are in contrast with those of Golden, Mortati, and Schroeter (1970) who reported a case of "burning feet" developed during L-DOPA therapy and a reduced plasma level of pyridoxine to 31 ng/ml. However, since their normal range for plasma pyridoxine varied from 35–80 ng/ml, it is questionable if this level constitutes a significant reduction.

III. DISCUSSION

It seems probable that the clinical improvements produced in parkinson patients by L-DOPA reflect a partial correction of the striatal dopamine deficiency. Thus, the finding that pyridoxine reduces the clinical response to L-DOPA was surprising since pyridoxine enhances the decarboxylation of DOPA to dopamine and would consequently be expected to augment the effect of exogenous L-DOPA. Our laboratory finding that a given dose of L-DOPA yields considerably lower plasma levels of DOPA when pyridoxine is administered concomitantly suggests an explanation for this apparent paradox. With lower plasma levels, the amount of L-DOPA available for transport across the blood-brain barrier and conversion to dopamine in the striatum must be greatly reduced.

How pyridoxine exerts this effect on the blood levels of DOPA remains uncertain. Altered absorption from the gastrointestinal tract appears unlikely because similar effects were noted whether pyridoxine was given parenterally or orally, and plasma HVA levels were not decreased by pyridoxine. We suspect that pyridoxine accelerates the decarboxylation and transamination of L-DOPA in the liver, gastrointestinal tract, kidney, and other organs, and hence reduces its availability for transport to the nervous system. Since the enzymes requiring pyridoxal phosphate are normally not fully saturated with their co-factor, enhanced activity may result when the daily intake of pyridoxine is increased. The possibility of formation of a pyridoxine-DOPA condensation product must also be considered. Such a compound forms readily *in vitro* (Golden, Mortati, and Schroeter, 1970; Holtz and Palm, 1964) and has recently been involved by Black and Axelrod (1969) as an explanation for their observation that liver transaminase levels are reduced after injection of norepinephrine. To date we have not been able to demonstrate a similar condensation between DOPA and pyridoxal phosphate in animal studies. Further evidence that pyridoxine negates the effects of DOPA by acting extracerebrally has been found by Cotzias (*personal communication*, 1971). In treating parkinson patients with a combination of L-DOPA and D,L-alpha-methyldopa hydrazine, Cotzias found no adverse effects from the addition of pyridoxine. We have recently confirmed these findings and additionally noted that in the presence of alpha-methyldopa hydrazine pyridoxine may actually enhance the clinical effects of levodopa (Yahr and Duvoisin 1971). In view of the fact that alpha-methyldopa hydrazine is a DOPA decarboxylase inhibitor which does not cross the blood-brain barrier, its effect in blocking the action of pyridoxine must take place in peripheral structures. Lacking such an action centrally allows for pyridoxine to accelerate the intracerebral metabolism of L-DOPA, presumably its decarboxylation and formation of dopamine.

Despite the apprehension expressed regarding the production of a pyridoxine-deficient state during L-DOPA therapy, there is no experimental evidence that avitaminosis is produced.

IV. SUMMARY

The use of pyridoxine hydrochloride as a therapeutic agent in parkinsonism has been evaluated. Administered by itself it does not ameliorate the symptoms of parkinsonism even in doses as high as 1000 mg per day. When pyridoxine hydrochloride is given to patients receiving L-DOPA, it nullifies its effects on the central nervous system. The beneficial clinical response as well as side effects such as abnormal involuntary movements are abolished even with doses as small as 5 mg of pyridoxine per day given orally. Evidence that this effect occurs by accelerating the decarboxylation and transamination of L-DOPA extracerebrally is presented. In patients on long-term L-DOPA treatment, neither clinical nor laboratory evidence of pyridoxine deficiency has been found.

REFERENCES

Andén, N., Roos, B., and Werdinius, B. (1963): On the occurrence of homovanillic acid in brain and cerebrospinal fluid and its determination by a fluorometric method. *Life Sciences, 7*:448–458.

Anton, A. H., and Sayre, D. F. (1962): A study of the factors affecting the aluminum oxide-tri-hydroxyindole procedure for the analysis of catecholamines. *Journal of Pharmacology and Experimental Therapeutics, 138*:360.

Antopol, W., and Schotland, C. E. (1940): The use of vitamin B_6 in pseudo-hypertrophic muscular dystrophy. *Journal of the American Medical Association, 114*:1058.

Baker, A. B. (1941): Treatment of paralysis agitans with vitamin B_6 (pyridoxine hydrochloride). *Journal of the American Medical Association, 116*:2484.

Barker, W. H., Stein, H. J., Miller, M. H., and Wintrobe, M. M. (1941): Failure of pyridoxine (vitamin B_6) to modify the parkinsonian syndrome. *Bulletin of Johns Hopkins Hospital, 69*:266.

Bertler, A., Carlsson, A., and Rosengren, E. (1958): A method for the fluorimetric determination of adrenaline and noradrenaline in tissues. *Acta Physiologica Scandinavica, 44*:273.

Black, I. B., and Axelrod, J. (1969): Inhibition of tyrosine transaminase activity by norepinephrine. *Journal of Biological Chemistry, 244*:6124.

Chang, C. C. (1964): A sensitive method for spectrophotofluorometric assay of catecholamines. *International Journal of Neuropharmacology, 3*:643.

Cotzias, G. C., Papavasiliou, P. S., and Gellene, R. (1969): L-DOPA in parkinson's syndrome. *New England Journal of Medicine, 281*:272.

Cotzias, G. C. and Papavasiliou, P. (1971): Blocking the negative effects of pyridoxine on patients receiving levodopa. *Journal of the American Medical Association, 215*:1504.

Ehringer, H., and Hornykiewicz, O. (1960): Verteilung von noradrenalin und dopamin (3-hydroxy-tyramin) im gehirn des menschen und ihr verhalten bei erkrankungen des extrapyramidalen system. *Klinische Wochenschrift 38*:1236.

Finke, H. (1954): Über die behandlung des parkinsonischen syndroms mit hohen dosen vitamin B_6. *Münchener Medizinische Wochenschrift, 96*:637.

Golden, R. L., Mortati, F. S., and Schroeter, G. A. (1970): Levodopa, pyridoxine and the burning feet syndrome. *Journal of the American Medical Association, 213*:628.

Hamfelt, A. (1967): Enzymatic determination of pyridoxal phosphate in plasma by decarboxylation of L-tyroxine-^{14}C(U) and a comparison with the tryptophan load test. *Scandinavian Journal of Clinical and Laboratory Investigation, 20*:1.

Hartmann-von Monakow, K. (1961): Die Bedeutung des vitamin B_6 für die neurologie. *Phychiatria et Neurologia (Basel), 142*:387.

Holtz, P., and Palm, D. (1964): Pharmacologic aspects of vitamin B_6. *Pharmacological Reviews, 16*:113.

Jameson, H. D. (1970): Pyridoxine for levodopa-induced dystonia. *Journal of the American Medical Association, 211*:1700.

Jollife, N. (1940): Effects of vitamin B_6 in paralysis agitans. *Transactions of the American Neurological Association, 66*:54.

Metzel, E., Weinmann, D., and Reichert, T. (1969): A study of the enzymes of dopa metabolism in parkinsonism from biopsies of the basal ganglia. In: *Third Symposium on Parkinson's Disease*, edited by F. J. Gillingham, and I. M. L. Donaldson, pp. 47–50. Livingstone, Edinburgh.

Sigwald, J. (1957): Les indications de la pyridoxine à Forte dose en neurologie. *Medicina Contemporanea, 75*:45.

Spies, T. D., Bean, W. G., and Ashe, W. F. (1939): A note on the use of vitamin B_6 in human nutrition. *Journal of the American Medical Association, 112*:2414.

Umbreit, W. W., Bellamy, W. D., and Gunsalus, I. C. (1945): The function of pyridoxine derivatives: A comparison of natural and synthetic codecarboxylase. *Archives of Biochemistry and Biophysics, 7*:185.

Yahr, M. D., Duvoisin, R. C., Schear, M. J., Barrett, R. E., and Hoehn, M. M. (1969): Treatment of parkinsonism with levodopa. *Archives of Neurology, 21*:343.

Yahr, M. D. (1970): Abnormal involuntary movements induced by L-DOPA. In: *L-DOPA and Parkinsonism,* edited by A. Barbeau and F. McDowell, pp. 101–108. F. A. Davis, Philadelphia.

Yahr, M. D. (1971): Pyridoxine, levodopa, and L-alpha-methyldopa hydrazine regimen in parkinsonism. *Journal of the American Medical Association, 216*:2141 (Letters).

Zeligs, M. A. (1941): Use of pyridoxine hydrochloride (vitamin B_6) in parkinsonism. *Journal of the American Medical Association, 116*:2148.

Zenker, N. (1966): The interaction of pyridoxal phosphate with tyrosine analogs. *Journal of Medicinal Chemistry, 9*:826.

Zimmerman, H. M. (1943) Pathology of vitamin B group deficiencies. In: *Proceedings, Association for Research in Nervous and Mental Disease,* Vol. 22, p. 51. Williams and Wilkins Co., Baltimore.

Advances in Biochemical Psychopharmacology, Vol. 4.
Raven Press, New York © 1972.

The Possible Role of Gamma-Aminobutyric Acid in the Synthesis of Protein

C. F. Baxter, S. Tewari*, and S. Raeburn

Neurochemistry Laboratories, Veterans Administration Hospital, Sepulveda, California;
Division of Neurosciences, City of Hope National Medical Center, Duarte, California;
and Department of Physiology, University of California, School of Medicine, Los Angeles, California

I. INTRODUCTORY BACKGROUND

A. Protein Metabolism and B_6 Vitamers

Six years ago, at another symposium on vitamin B_6, Axelrod and Trakatellis (1964) drew attention to the importance of vitamin B_6 for the immune response in animals and demonstrated inhibition of the *in vivo* incorporation of ^{14}C-valine into the proteins of liver and spleen in vitamin B_6-deficient rats (Trakatellis and Axelrod, 1964). These observations stimulated our interest and led to some preliminary experiments to test the effect of vitamin B_6 deficiency on the protein-synthesizing capacity of brain tissues in weanling rats.

Three groups of rat pups were suckled by dams on different diets: a commercial standard diet (Group 1), a synthetic diet[1] (Group 2), and a synthetic, vitamin B_6-deficient diet[1] (Group 3). The standard diet consisted of Purina Rat Chow. The synthetic diet contained 20% GBI vitamin-free casein, 68% starch, 5% hydrogenated vegetable oil, 4% USP XIV salt mixture, 1% choline bitartarate, and the remaining 2% as a normal vitamin supplement. For the

---

* Present address: Department of Psychiatry, University of California, Irvine, California.

[1] The synthetic diets were formulated for Berkeley Pacific Laboratories by Dr. M. A. Williams, University of California at Berkeley.

third group, vitamin B_6 was omitted from the vitamin supplement. The dietary regimen was begun 3 weeks prior to parturition and continued throughout the lactation period. All diets were fed *ad libitum.* At the time of weaning, the protein-synthesizing capacity of brain and liver tissues from the rat pups was examined. The incorporation of ^{14}C-leucine into protein was measured in cell-free *in vitro* systems, as described previously (Tewari and Baxter, 1969) and in Section II of this manuscript.

Results of this study were in full agreement with the earlier *in vivo* data. When organs from pups nursed by dams on a synthetic B_6-deficient diet were compared with organs from pups nursed by dams on a synthetic diet, the protein-synthesizing capacity of brain and liver of the B_6-deficient pups appeared depressed by 21% and 40%, respectively. However, the most notable and unexpected result of the study was a clear indication that the synthetic diet, which appeared adequate for the dams (on the basis of body weight), produced severely deficient pups in the absence and in the presence of pyridoxine. Thus, *all* pups from dams on a commercial standard diet (Group 1) weighed an average of 45 g at weaning time, pups from dams on the synthetic diet (Group 2) weighed an average of only 25 g, and pups from vitamin B_6-deficient dams on the synthetic diet (Group 3) weighed an average of 17 g. The retarded and weak condition of the vitamin B_6-deficient pups was expected (Davis, Nelson, and Shepard, 1970), but the great discrepancy in the body weight between the pups from Groups 1 and 2 indicated an inadequate milk flow, in quality or quantity, from dams on the synthetic diet.

A decreased food intake is a concomitant of many vitamin B_6-deficient synthetic diets, and inanition is known to decrease both levels of protein and rates of protein synthesis in a variety of organs (Lee, 1970; Fleck, Shepherd, and Munro, 1965; see also Munro, 1964, for other *in vivo* examples). In weanling rats, protein synthetic mechanisms are particularly susceptible to short periods of inanition. When 23-day-old Sprague-Dawley rats were deprived of their regular lab chow for one day, their body weights declined by 10 to 20%. Protein-synthesizing capacity of their brain tissues declined an average of 24% when expressed as $\mu\mu$moles U[^{14}C]leucine incorporated into protein/mg ribosomal protein/hr, and an average of 38% when calculated for the total tissue of each rat brain. The difference in the two percentages reflects the rapid loss of ribosomal protein from brain tissue of the starved rats. These results (Tewari and Baxter, *unpublished*) are compatible with observations by Zamenhof, Marthens, and Margolis (1968) who demonstrated under similar experimental conditions that both the absolute protein content and the cell numbers decrease in the neonatal rat brain.

All of the foregoing experimental data suggest that both an inadequate food intake and a nutritional vitamin B_6 deficiency lead to an impaired rate of protein synthesis in brain tissues of weanling rats. They offer little insight into the mechanisms by which vitamin B_6 vitamers might regulate protein synthesis. Substrate levels of some amino acids are known to affect the rate of protein metabolism (Munro, 1968; Tewari and Baxter, 1969), and vitamin B_6 is the established cofactor for many enzymatic reactions involving amino acid synthesis and degradation. It is plausible, therefore, to attribute the effect of a vitamin B_6-deficient diet to the inhibition of some B_6-dependent decarboxylation or transamination, provided that the level of the amino acid metabolized in such a reaction is rate-limiting in the synthesis of one or more proteins.

Alternately, the effect of a vitamin B_6 deficiency on protein synthesis could be mediated by any compound which modifies, directly or indirectly, the rate of protein metabolism, as long as the compound is itself regulated by vitamin B_6-dependent enzyme systems. It is this latter possibility that relates γ-aminobutyric acid (GABA) and protein synthesis to the overall theme of this symposium on B_6 vitamers.

B. γ-Aminobutyric Acid and Vitamin B_6

GABA is both synthesized and degraded by enzymes requiring pyridoxal phosphate as a cofactor. In tissues of the nervous system, levels of GABA are more sensitive than those of most other amino acids to pharmacologically induced vitamin B_6 deficiencies and to vitamin B_6 antagonists (see Baxter, 1970, for review). Experimental results to be presented in this paper support the concept that GABA is a potential regulator of protein metabolism in the brain tissues of the weanling rat.

C. Protein Metabolism and Neurotransmitter Agents

The regulation of protein metabolism in the nervous system is of particular interest because of the popular belief that this process is correlated with memory and learning (see Hydén and Lange, 1970, for references). Many chemical entities regulate directly or indirectly the rate of protein synthesis. Campbell, Mahler, Moore, and Tewari (1966) drew attention to the possibility that compounds with physiological activity at synaptic junctions (neurotransmitters) also might have a regulatory role in the synthesis of proteins of nervous tissues. GABA was noteworthy among these compounds in that it *stimulated* the incorporation of amino acids into protein.

This challenging observation, which pointed to the possible interrelationship between some mediators of synaptic transmission and protein synthesis, led us to initiate experiments which confirmed and expanded the original observation (Tewari and Baxter, 1969; Baxter and Tewari, 1970). Of more than 20 amino acids that were tested at physiological concentrations, only glycine and GABA clearly stimulated the incorporation of ^{14}C-leucine into proteins of a cell-free system derived from immature rat brain. Slight effects were noted also with glutamic acid and histidine. Perhaps it is not just coincidental that all of the aforementioned amino acids have some physiological properties of neurotransmitter agents.

Several neurotransmitters have recently been related to protein metabolism. Protein synthesis in the parotid gland is stimulated by epinephrine (Grand and Gross, 1970), and the incorporation of ^{14}C-tryptophan into protein in pineal gland tissue culture is stimulated by norepinephrine (Wurtman, Shein, Axelrod, and Laren, 1969). An unidentified agent in nerve fibers has been implicated as a stimulator of protein synthesis in the newt (*Triturus*) (Lebowitz and Singer, 1970).

Attempts at correlating electrical activity and stimulation of nervous tissues with the synthesis or turnover of proteins has produced contradictory results. The reports to the effect that electrical stimulation (which presumably releases neurotransmitters) is accompanied by an increased metabolism of protein and RNA (Singh and Talwar, 1967; Berry, 1969; Luxoro, 1960; Appel, Davis, and Scott, 1967) are opposed by reports which appear to show exactly the opposite (Orrego and Lipmann, 1967; Metzger, Cuenod, Grynbaum, and Waelsch, 1966). Nevertheless, based on the positive evidence, the concept that some neurotransmitters or neuromodulators can affect protein metabolism, as do hormones and polyamines, is gaining acceptance.

II. METHODOLOGY FOR THE IN VITRO SYSTEM

The cell-free incubation system used in our initial experiments has been described in detail (Tewari and Baxter, 1969). Briefly, it contains a ribosomal and a pH 5 enzymes fraction derived from 21-day-old rat brains. For incubation at 37°C, these fractions are suspended in a solution containing Trizma buffer (pH 7.6), MgAc, NaCl, KCl, mercaptoethanol, sucrose, ATP, GTP and ^{14}C-labeled amino acid. In the incubation mixture, the ratio of ribosomal protein to pH 5 enzymes protein varied from 1:3.0 to 1:3.5. The incorporation of ^{14}C-labeled amino acids into the hot trichloracetic acid (TCA) precipitable protein pellets served as an index of protein-synthesizing activity. Radioactivity incorporated into aminoacyl-tRNA was determined in early experiments by measuring the difference in counts between pellets extracted with hot TCA (protein) and pellets extracted with ice cold TCA (aminoacyl-tRNA and protein). When ribosome-free incubation systems were used, no appreciable amount of radioactivity was found in the hot TCA precipitable fraction, and measurements of aminoacyl-tRNA could be made directly by measuring incorporation into the cold TCA precipitable fraction.

U[^{14}C]-amino acid incorporation into the aminoacyl adenylate enzyme complex (amino acid activation) was measured by the formation of amino acid hydroxamate (Loftfield and Eigner, 1963). The free hydroxylamine base present during the course of the reaction was freshly prepared from hydroxylamine hydrochloride by the method of Beinert, Green, Hele, von Korff, and Ramakrishnan (1963). The resolution of U[^{14}C]leucyl hydroxamate from free U[^{14}C]leucine was carried out on Reeves Angel SA 2 ion-exchange paper.

The U[^{14}C]leucyl-adenylate-enzyme complex was synthesized in a system containing 1.52 mg aminoacyl-tRNA synthetase free of tRNA's, 16 μC of U[^{14}C]leucine, 10 μmoles ATP, and 1.4 μmoles Mg^{++}, all in a final volume of 1 ml of medium M. Incubation of this system for 10 min at 37°C and immediate inactivation with 0.05 ml of 0.1 M EDTA (Lagerkvist, Rymo, and Waldenström, 1966) yielded ^{14}C-leucyl-adenylate-enzyme complex which was isolated on a Sephadex G-75 column (Rouget and Chapeville, 1968). The tRNA's and aminoacyl-tRNA synthetases (free of tRNA's) were isolated from a specially prepared pH 5 fraction, as shown in Fig. 1. The scheme is based on a combination of methods described by Wong and Mustard (1969) and by

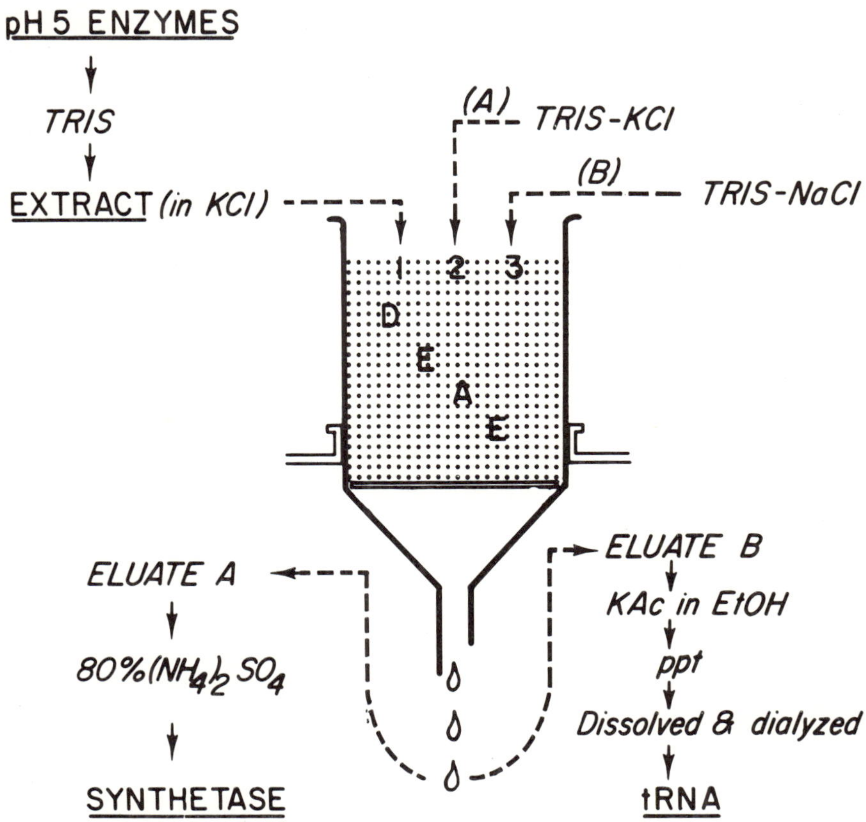

FIG. 1. Scheme for the separation and recovery of tRNA and aminoacyl-tRNA synthetase from brain tissues on a DEAE column.

Takeishi, Nishimura, and Ukita (1967). The pH 5 fraction was prepared as follows. The post-microsomal supernatant (cell sap) was diluted with two volumes of ice-cold distilled water and then adjusted to pH 5.2 with 1 N HAc. It was spun at 12,800 *g* for 16 min. The supernatant was discarded. The precipitate was dissolved as far as possible in 50 mM Trizma buffer, pH 7.5, and undissolved material was separated by centrifugation. The supernatant contained most of the synthetases and tRNA's. The precipitate was washed twice with 50 mM Trizma buffer and the extracts pooled with the original supernatant. To this extract solid KCl was added to produce a final concentration of 0.33 M. The extract was applied to a DEAE cellulose column. Elution with Tris-KCl yielded the synthetases; further elution with Tris-NaCl yielded a fraction containing the tRNA's. For details of the method, the reader should consult the aforementioned references.

U[^{14}C]leucyl-tRNA was extracted from the pH 5 fraction (after incubation) by the method of Hoagland, Stephenson, Scott, Hecht, and Zamecnik (1958) and further purified by a slight modification of the method of Bloemendal, Littauer, and Daniel (1961). The modification consisted of the addition of 1% Na dodecyl sulfate to the phenol. This improved the extraction of the tRNA. Other methodological considerations will be found in the legends to specific figures and tables.

All counting of radioactivity was performed with an ANSitron liquid scintillation counter, as described previously (Tewari and Baxter, 1969). Results expressed as counts per minute have been corrected for background and occasional quenching. The counting efficiency ranged from 76 to 80% during the years 1966 to 1970.

III. PROPERTIES OF THE IN VITRO SYSTEM

A. The Effect of GABA on the Ribosomal System

Our earlier studies have shown that 4mM GABA added to a cell-free ribosomal system from immature rat brain stimulated the incorporation of U[^{14}C]leucine into proteins by 40 to 50%. This stimulation was fairly unique; no GABA analog or any ω-amino acid had a similar effect. GABA stimulated the incorporation of several amino acids into protein but to different degrees. The stimulatory effect of GABA was abolished in the absence of Na^+ in the incubation mixture and in the presence of some amino acids (including glycine). However, low to moderate levels of amino acid mixtures added exogenously to the incubation mixture did not impair the GABA effect. GABA was not degraded to any measurable extent in these incubation mixtures.

B. Experiments Pertaining to the Mechanism of GABA Action

Incomplete data suggested but did not prove that GABA was not inhibiting proteinase activity or RNAse activity, and did not contribute to the energy metabolism of the *in vitro* system. No evidence was found that GABA affected protein metabolism by its degradation to ammonia, as suggested by Gershenovich, Krichevskaya, and Shumskaya (1965). In contrast to GABA, the complete *in vitro* ribosomal system was not stimulated by up to 5mM NH_4Cl. (However, a purified pH 5 enzymes system, studied at a later date, can be affected by NH_4Cl.)

The GABA stimulation of leucine incorporation into protein was not Mg^{++} dependent. Whereas the absolute stimulation by GABA ($\mu\mu$moles leucine incorporated/mg ribosomal protein) remained constant over a wide

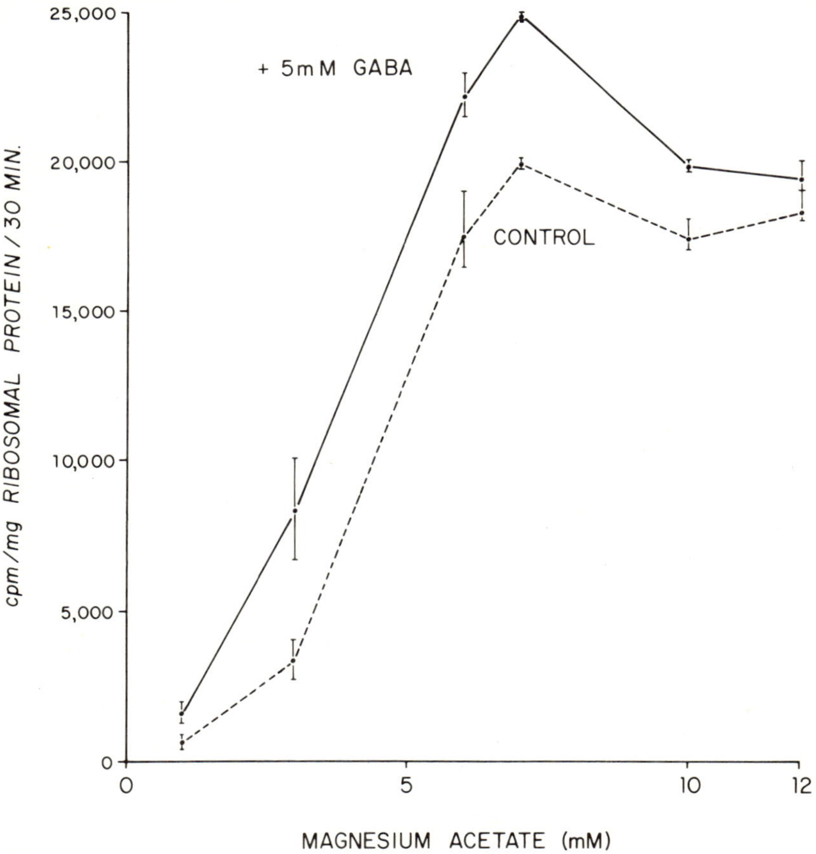

FIG. 2. Effect of magnesium acetate concentration on the incorporation of U[^{14}C]leucine into proteins of ribosomal system from brain tissues. The vertical bars represent the range of values in three experiments.

range, the relative stimulation by GABA (% of total activity) decreased at higher Mg^{++} concentrations. At low Mg^{++} concentrations (1 to 3 mM), which more closely resemble *in vivo* levels (Bachelard and Goldfarb, 1969), the stimulation of leucine incorporation by GABA exceeded 100% (Fig. 2). The GABA effect may, therefore, be much more pronounced *in vivo* than in the usual *in vitro* system with a 7 to 10 mM Mg^{++} concentration.

The GABA effect was dependent on the concentration of the pH 5 enzymes in the incubation mixture, or, more precisely, on the ratio of pH 5 enzymes to ribosomes (Fig. 3). All experiments with the ribosomal system were run with close to an optimal ratio between the two fractions.

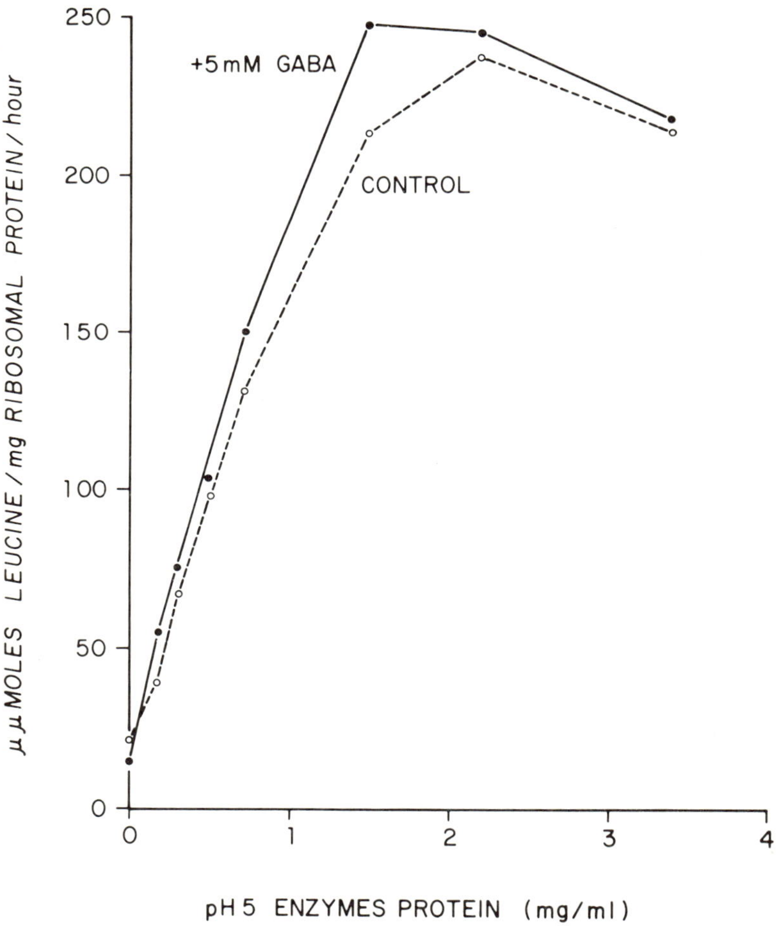

FIG. 3. Effect of GABA and pH 5 enzymes concentration upon U[^{14}C]leucine incorporation into proteins of ribosomal system from brain tissues. Ribosomal protein (0.456 mg) was present in the 1 ml incubation mixture.

C. Organ Specificity of the GABA Effect

The search for the site of GABA action in the regulation of protein synthesis in cerebral tissues was aided unexpectedly by a comparative study. The effect of GABA was studied in cell-free, protein-synthesizing systems from brain, liver, kidney, heart, and reticulocytes of 21-day-old rats. Ribosomes and pH 5 enzymes were prepared from each tissue by similar procedures, except that heart ribosomes were first isolated in a medium low in Na^+ and K^+, and reticulocyte preparations were not subjected to desoxycholate treatment. The purity of different ribosomal preparations was verified by electron microscopy and their size by analytical ultracentrifugation. Only the ribosomes of the reticulocyte preparation were contaminated with appreciable amounts of membranous material.

In cell-free leucine-incorporating systems from liver, heart, and reticulocytes, the addition of GABA had no stimulatory effect upon the incorporation of U[^{14}C]leucine into protein. The GABA effect was confined to preparations from brain and, to a lesser extent, kidney (Baxter and Tewari, 1970); both organs contain appreciable quantities of endogenous GABA in their tissues (Zachman, Tocci, and Nyham, 1966).

The relative organ specificity of GABA stimulation was used as a tool to investigate the site of GABA action. Results of experiments in which brain ribosomes were mixed with pH 5 enzymes from brain, liver, kidney, heart, and reticulocytes are shown in Table 1. Only the pH 5 enzymes from brain were compatible with the GABA effect. Conversely, if pH 5 enzymes from brain were mixed with ribosomes from brain, liver, heart, and reticulocytes, the addition of GABA to the incubation mixture stimulated leucine incorporation

TABLE 1. *Effect of GABA on U[^{14}C] leucine incorporation into proteins of system containing brain ribosomes and a pH 5 enzymes fraction from different organs*

Source of		Activity ($\mu\mu$moles leucine incorp./mg prot./hr)	
Ribosomes	pH 5 Enzymes fraction	control	+ 5mM GABA
brain	brain	88	111
	liver	77	76
	kidney	42	40
	reticulocytes	35	34

The results shown are the averages of three experiments. Although the absolute $\mu\mu$moles of leucine incorporated/mg protein/hr varied from one preparation to the next, percentage changes observed with the addition of GABA were consistent in every experiment ($\pm$ 5%).

in all of these preparations. In kidney ribosomes, the addition of brain pH 5 enzymes fraction alone increased leucine incorporation by 50 to 100% above that found with the kidney pH 5 fraction. The addition of GABA had no further significant effect (Table 2).

TABLE 2. *The effect of GABA on U[^{14}C] leucine incorporation into proteins. Ribosomal cell-free systems from different organs and substitution of brain pH 5 enzymes fraction for similar fraction from other organs*

Source of: Ribosomes	Source of: pH 5 Enzymes fraction	Activity (μμmoles leucine incorp./mg prot./hr): Control	Activity (μμmoles leucine incorp./mg prot./hr): + 5mM GABA
brain	+ brain	88	111
liver	+ liver	65	
	+ brain	89	110
kidney	+ kidney	22	
	+ brain	63	59
reticulocytes	+ reticulocytes	8	
	+ brain	19	23

The incubation time was 30 min. The results shown are the averages of three experiments. Although the absolute μμmoles of leucine incorporated/mg protein/hr varied from one preparation to the next, the percentage changes observed with the addition of GABA were consistent for brain and liver within ± 5%. When the pH 5 fraction of kidney and reticulocyte preparations was replaced with the pH 5 fraction from brain, the increased level of leucine incorporation varied widely (± 25%). However, in both preparations, the addition of GABA had no significant effect in any of the experiments.

D. Identification of a Specific Step in Protein Synthesis Stimulated by GABA

The foregoing experiments indicated that specificity for the GABA effect appeared to be in the pH 5 enzymes fraction of the brain tissues and not in the ribosomal fraction. To verify this conclusion, U[^{14}C]leucine was incubated with the pH 5 enzymes fraction alone and the ^{14}C-leucyl-tRNA precipitated and counted as described under Methodology. The addition of GABA stimulated the formation of leucyl-tRNA as shown in Fig. 4. The corollary experiment in which some ^{14}C-leucyl-tRNA was first isolated and then incubated with ribosomes, GTP, and some pH 5 enzymes (as a source of transferases) is shown in Fig. 5. GABA had no apparent effect either on the incorporation rate of the ^{14}C label into protein or on its excorporation out of the ^{14}C-leucyl-tRNA complex. Since the amount of ^{14}C-leucyl-tRNA substrate may have been

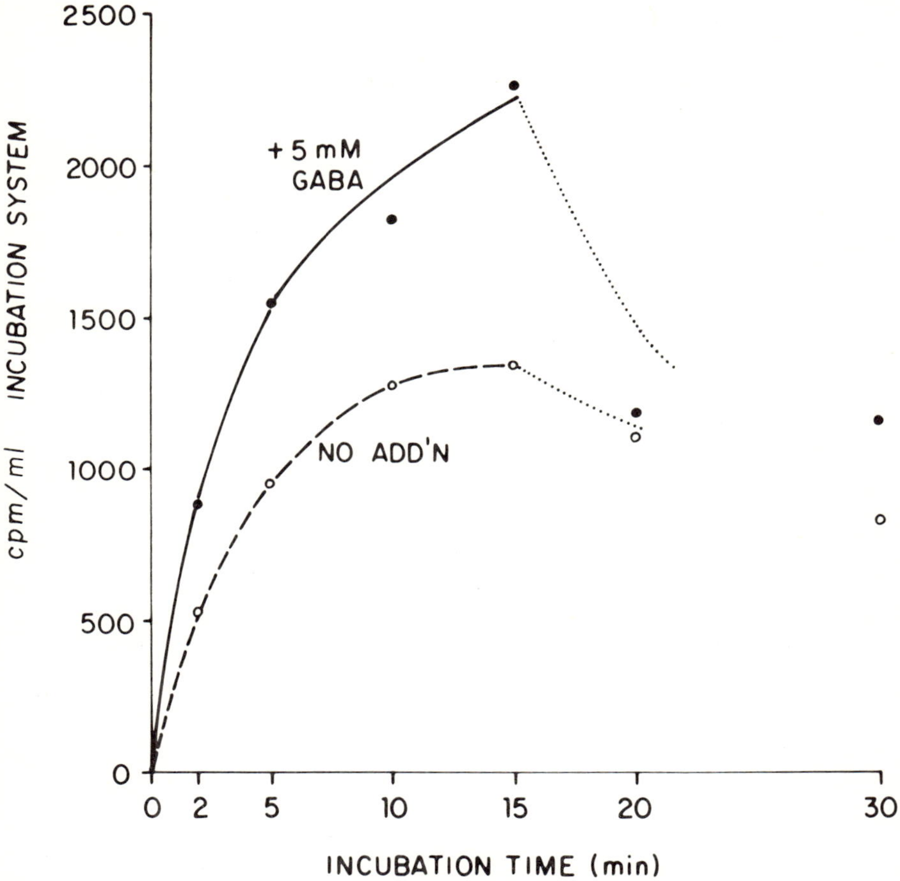

FIG. 4. Effect of GABA upon the rate of U[^{14}C]leucyl-tRNA formation in cell-free system from rat brain.

rate-limiting, this experiment *per se* was considered inconclusive. However, in conjunction with the organ specificity of the pH 5 fraction and the stimulation of leucyl-tRNA formation by GABA, the evidence was sufficient to permit the abandonment of studies on the ribosomal fraction and to concentrate on the pH 5 enzyme fraction. Aside from the transfer factors mentioned above, this fraction contained synthetases, tRNA's, and, presumably, initiation and termination factors.

The stimulation of aminoacyl-tRNA by GABA was found to be independent of all but very low concentrations of ATP (Fig. 6). Since ATP is an absolute requirement of the amino acid activation step for protein synthesis, this first step was tested next, using the amino-acid adenylate-enzyme complex trapping technique in which the hydroxamate is formed and isolated (see

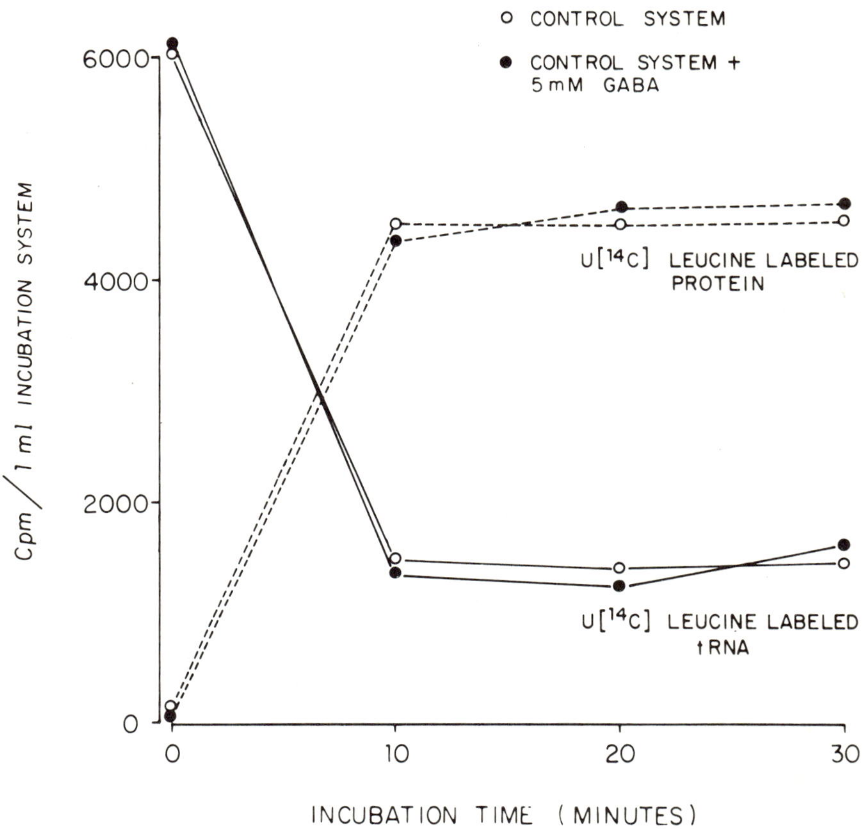

FIG. 5. Effect of GABA upon the rate of transfer of U[^{14}C]leucine from U[^{14}C]leucyl-tRNA into the protein of the ribosomal system. 350 μg of U[^{14}C]leucyl-tRNA containing 6,300 cpm (extracted with phenol containing dodecyl sulfate) was added to an incubation system containing 1.48 mg of ribosomal protein and 4.49 mg of pH 5 enzymes protein. The latter was added to provide a source of transfer enzymes. Final volume of system was 5 ml.

Methodology). The results shown in Fig. 7 verified that this step was not stimulated by GABA.

To prove that GABA acted on the one remaining step, it was necessary to isolate an appreciable amount of the ^{14}C-leucyl-adenylate-enzyme complex in reasonably pure form from brain, as well as cerebral tRNA. Initial efforts were unsuccessful because the aminoacyl-adenylate-enzyme complex from rat brain proved to be quite unstable. In subsequent experiments, it was possible to isolate some of the labeled complex and this was used in four experiments, as shown in Table 3. Although the absolute results differed considerably from one experiment to another and the radioactivity incorporated

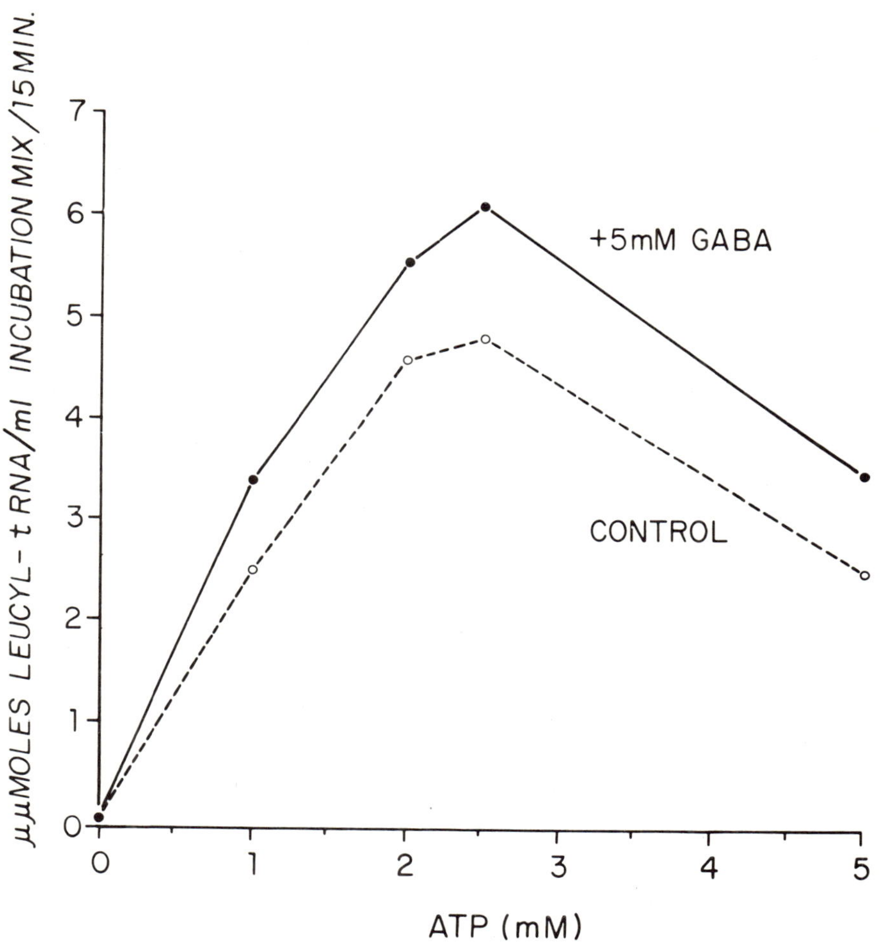

FIG. 6. Effect of ATP and GABA on the rate of U[^{14}C]leucyl-tRNA formation in cell-free system from brain. The 1 ml incubation mixture contained 1.12 mg of pH 5 enzymes protein and a Mg^{++} concentration of 7 mM. Incubations for 15 min at 37°C.

into the end product was low, GABA showed a consistent stimulatory effect upon the formation of ^{14}C-leucyl-tRNA.

Obviously, the technique for the preparation of the ^{14}C-leucyl-adenylate-enzyme complex from brain still requires considerable refinement, but it appeared virtually certain that one reaction step in the synthesis of protein, which was affected by GABA, had been identified. It remained to be shown whether the specificity for GABA resided in the tRNA or the synthetase enzyme.

This point was investigated by first preparing a purified synthetase and a purified tRNA fraction from immature rat brain and liver. Using organ

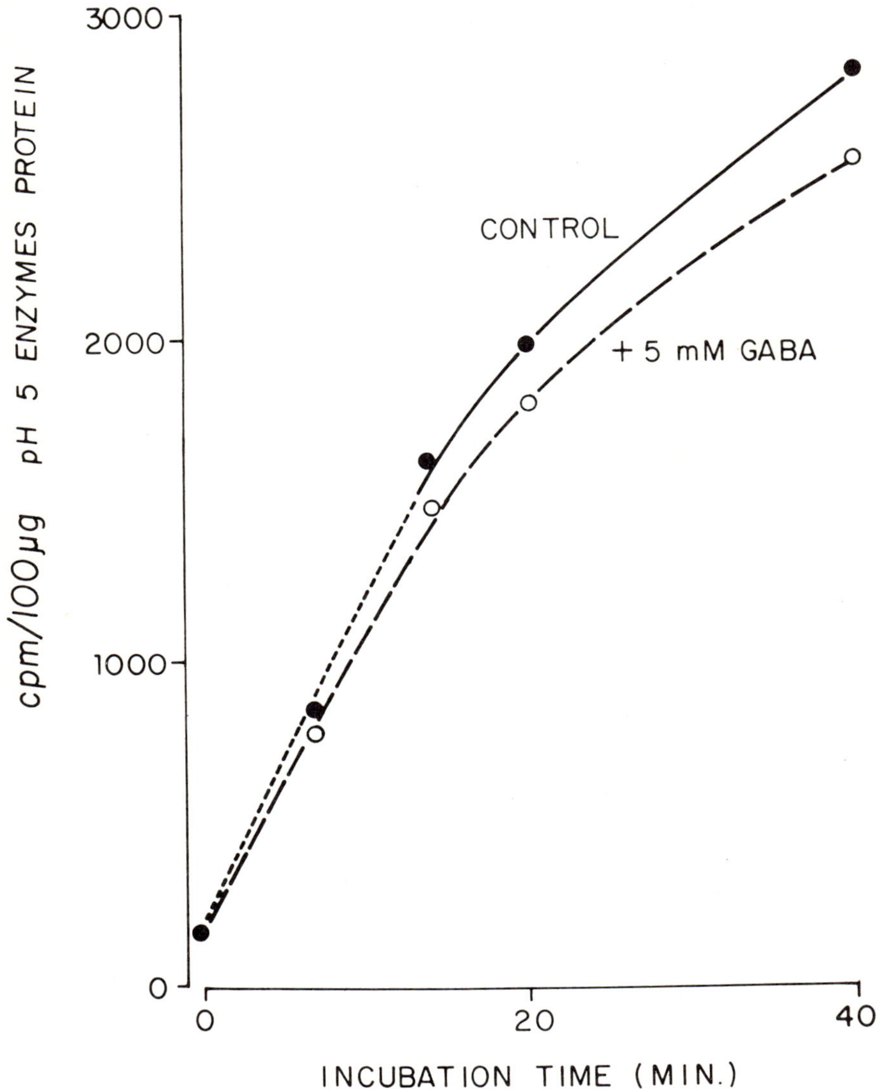

FIG. 7. The effect of GABA on the rate of leucine activation in cell-free system from immature rat brain.

specificity once again as a tool, it was possible to show in cross-over experiments that brain synthetase with liver tRNA could be stimulated by GABA to increase incorporation of ^{14}C-leucine into the leucyl-tRNA complex. However, GABA had no such effect on a system containing liver synthetase and brain tRNA. The complete but preliminary data are shown in Fig. 8,

TABLE 3. *Formation of leucyl-tRNA from leucyl-adenylate-enzyme complex and tRNA in immature rat brain (synthetase—2nd step)*

Incubation conditions	Experiments			
	1	2	3	4
μg protein	63	68	95	40
μg tRNA	13	13	13	13
	cpm/100μg tRNA*			
control	516	338	1517	909
+5mM GABA	986	423	2033	1509
percent stimulation by GABA	91%	25%	33%	66%

*All incubations were for 10 min only at 37°C and in the presence of 7 mM Mg^{++} in the 1-ml system of Medium M. The ^{14}C-leucyl-adenylate-enzyme complex was synthesized by incubating 4.98 mg of aminoacyl-tRNA synthetase enzyme in 4 ml of an incubation mixture containing 40 μmoles Tris-ATP, 4.2 μmoles Mg^{++}, 0.73 μmoles cold leucine, and 10 μC U[^{14}C]leucine. This mixture was incubated for 10 min at 37°C and the reaction stopped by cooling to 4°C. The ^{14}C-leucyl-adenylate-enzyme complex was isolated on an G-75 Sephadex column in the cold.

	BRAIN SYNTHETASE	LIVER SYNTHETASE
BRAIN tRNA	+	—
LIVER tRNA	+	—

FIG. 8. Stimulation of U[^{14}C]leucyl-tRNA formation by GABA in purified systems with aminoacyl-tRNA synthetases and tRNA's from brain or liver.

and constitute data which were obtained prior to 10 months ago. They were interpreted to indicate that GABA interacted in some way with the synthetase.

III. AMINOACYL-tRNA SYNTHETASE AS THE TARGET FOR GABA INTERACTION

One point at which GABA interacts with the reaction sequence of protein synthesis has been identified, as shown in Fig. 9. This does not exclude the possibility that GABA might interact at other points in the reaction sequence *in vivo*. The preparation of the *in vitro* cell-free incubation system is such that many factors for protein synthesis may be lost or diluted in the process; consequently, the rate-limiting step *in vitro* may be entirely different from the rate-limiting step *in vivo*.

It remains unknown if the aminoacyl-tRNA synthetases—Step 2—can be rate-limiting in cerebral protein synthesis, but some evidence points in that direction. Some synthetases have been shown to be the targets of several factors which are believed to exert a regulatory function on protein synthesis. In *E. coli*, the ammonium ion has a stimulatory effect on leucyl-tRNA synthetase (Yu and Hirsh, 1967). Such an effect can be demonstrated also in the purified system from rat brain. Similarly, polyamines such as spermidine,

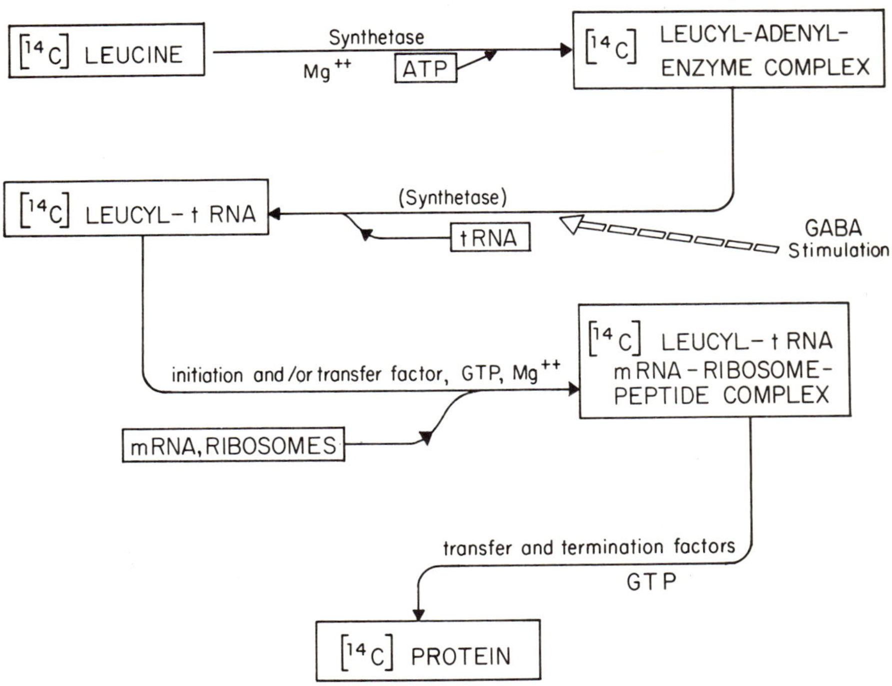

FIG. 9. An abbreviated scheme for the incorporation of amino acids into protein. Point of GABA regulation.

spermine, and putrescine stimulate aminoacyl-tRNA formation *in vitro*, provided the Mg^{++} concentration is not too high (Takeda and Igarashi, 1969). A number of polynucleotides also are known to inhibit the activity of aminoacyl-tRNA synthetases (Deutscher, 1968).

Specific aminoacyl-tRNA synthetases are affected by thyroxine (Yang and Samadi, 1969); the reaction sequence is inhibited by ADP and AMP in *Salmonella typhimurium* (Brenner, DeLorenzo, and Ames, 1970) and by aminoacyl adenosine in reticulocytes (Harbon and Chapeville, 1970). It is apparent also that salt concentrations, sulfhydryl reagents, and many other extraneous factors will alter the characteristics of the reaction sequence *in vitro* (Ritter, Kull, and Jacobsen, 1970; Smith, 1969; Manchester, 1970; Allende, Mora, Gatica, and Allende, 1965).

The stimulatory effect of GABA on the synthetase cannot be related to its own activation by synthetase and ATP. Although such a reaction is feasible, its occurrence would make GABA a competitor for synthetase, ATP, and possibly tRNA, and would tend to inhibit rather than stimulate the rate of protein synthesis. Consequently, the most likely mechanism by which GABA could exert its stimulatory effect would be by inhibiting an inhibitor in the incubation system. Since such an inhibitor could be extrinsic or intrinsic in nature, the physiological potential of the stimulatory effect of GABA on protein synthesis *in vivo* cannot be evaluated until the mechanism of the GABA effect is understood.

IV. RECENT PROBLEMS IN ASSESSING THE ROLE OF GABA IN THE IN VITRO SYSTEM

From 1966 through 1969, the stimulatory effect of GABA upon protein synthesis was observed quite consistently in hundreds of experiments. Early in 1970, however, when we were studying the time relationship of the GABA effect with respect to a variety of parameters, the stimulatory action of GABA in the *in vitro* system disappeared. Since the source of GABA was the same for all experiments and GABA was not metabolized in the system (see Section III, A), we assumed at first that the effect was lost because of the extensive purification of substrate and enzyme fractions. However, on retesting ribosomal and pH 5 fractions, as used for the early experiments in 1966, we found that even in this preparation GABA no longer stimulated leucine incorporation into protein. Several other characteristics of the system which had been observed consistently and repeatedly for 3–1/2 years previously were found to have changed. Something had happened that either affected the experimental animals, the *in vitro* system, or the methodology of the experimental procedures. Table 4 lists some of the experiments which have been conducted so far to discover the reason for the lack of stimulation by GABA.

TABLE 4. *Ribosomal protein synthesizing system from rat brain: variables investigated in the disappearance of the stimulatory effect by GABA*

I. Animals	*III. Preparation*
1. age	12. storage
2. diet	13. glassware and dishwashing
3. source	14. homogenization
4. bedding	15. precipitation of pH 5 enzymes
5. handling	
II. Chemicals	*IV. Incubation*
6. H_2O	16. glassware and dishwashing
7. Trizma buffer	17. incubation time
8. $MgAc_2$, $MgCl_2$	18. Mg^{++} levels
9. sucrose	19. pH 5/ribosome ratios
10. GABA	20. TCA precipitation of proteins
11. HAc	

A. Changes in Animals and Environment

Possible changes in diet, age, and environment of the experimental rats and their dams were investigated. A change in the bedding of the rat cages from cedar shavings to pine shavings coincided in time with the lost activity, but proved to be one of several changes which were unrelated to the GABA effect. One discrepancy of general interest is described below.

All weanling rats for our experiments were bred and received from Simonsen Laboratories (Gilroy, California). For years, this caviary has bred its own strain of Sprague-Dawley rats. In order to produce a "better" rat containing fewer pathogens, the caviary improved its strain of Sprague-Dawley rats toward the end of 1969 and early in 1970 as follows. Pregnant gnotobiotic females were purchased and placed into a brand new, specially cleaned and ventilated building. The pups of the gnotobiotic rats were destroyed and the lactating females were used as foster mothers for litters of Sprague-Dawley pups, which were delivered by Caesarian section from Simonsen's own rat colony. These rats were then used as the foundation for a new colony. In this way, the caviary had improved its stock and was delivering to its customers a "healthier and cleaner" rat with fewer pathogens. To our knowledge, few if any of Simonsen's many customers knew of this improvement. We heard of it only *after* the old colony had been completely eliminated, and we may never know whether the pathogens in the old colony had any effects upon our results in the past. However, experimental results with weanling rats from other caviaries led us to believe that this variable was not the crucial factor in changing the *in vitro* protein-synthesizing system from brain tissues.

B. Changes in the Reagents of the In Vitro Incubating System

If the mechanism of GABA action on the tRNA synthetase enzyme involves the inhibition of an inhibitor (disinhibition of the enzyme), then the loss of the GABA effect could indicate the loss of the inhibitor in the *in vitro* system. Since the origin of an inhibitor could be extrinsic, all chemicals used in the preparation and incubation of the leucine-incorporating system, as well as reagents used in washing laboratory glassware, were checked both by comparison with other lots of the same chemical and by sending letters of inquiry to the manufacturers. Some of this work is still in progress. Changes discovered in one of the many chemicals investigated is of general interest and is detailed below.

Trizma buffer, purchased from Sigma Chemical Company, was one of the first reagents checked. Mahler reported in 1969 that his laboratory had lost the stimulatory effect of GABA upon protein synthesis, but recovered the effect when they changed to a Tris buffer system (Mahler, 1970). A routine inquiry to Sigma revealed that Trizma manufactured early in 1969 was crystallized from a methanolic solution, whereas Trizma made late in 1969 and in 1970 was crystallized from aqueous solutions. We compared an old and a new batch of Trizma and found that, in our system, the old buffer (presumably crystallized out of methanol) supported a higher rate of leucine incorporation into protein, but neither was instrumental in restoring the GABA effect.

C. Changes in Experimental Techniques

More recently, attention has been focused on the many potential variables in the preparation of the ribosomes and pH 5 enzymes fraction prior to incubation. A detailed review of service records revealed that, close to the time that the stimulatory effect of GABA disappeared, the Radiometer pH meter used in these studies was recalibrated during a routine maintenance call, the drive shaft of the Spinco L-2 ultracentrifuge was replaced, and repairs were made to the storage tank of our centralized distilled-water system. Effects of these and other parameters which might have altered the experimental conditions are under investigation.

V. SUMMARY AND CONCLUSIONS

The rate of leucine incorporation into the proteins of brain tissue from weanling rats was measured in an *in vitro* cell-free system. In full agreement with earlier *in vivo* studies, both a nutritional vitamin B_6 deficiency and food deprivation decreased the rates of protein synthesis as measured by leucine incorporation into brain proteins of the *in vitro* system. GABA, a compound both synthesized and degraded by enzymes requiring pyridoxal phosphate as cofactors, stimulated the incorporation of amino acids into brain proteins in a cell-free system. This effect was independent of ATP or Mg^{++} concentrations and optimal at 4 to 5 mM GABA concentration. The stimulation of leucine incorporation was observed only in brain and kidney tissues; both of these organs contain endogenous GABA.

On examining individual reaction sequences along the pathway of protein synthesis, it was found that GABA stimulated selectively aminoacyl-tRNA synthetase activity as tested with U[^{14}C]leucine.

The recent loss of our ability to demonstrate the stimulatory activity by GABA upon the overall rate of leucine incorporation into the *in vitro* system may reflect the sensitivity of aminoacyl-tRNA synthetase to regulation by many extrinsic and intrinsic factors in the system. Present studies are designed to recover that ability and to gain insight into the *mechanisms* by which GABA stimulates protein synthesis.

Acknowledgments. The technical help of Mrs. B. Manely, Mr. J. Bright and Mr. R. Wilton and the secretarial help of Miss L. Eaton is gratefully acknowledged. This project was supported in part by NIH Grant No. NS 03743 and by a stipend to R. Wilton from the Spinal Cord Research Foundation.

REFERENCES

Akino, T., and Amano, M. (1970): Stimulation of amino acid incorporation into protein *in vitro* by various nuclear RNA fractions of the AH-130 hepatoma cells. *Journal of Biochemistry (Japan), 67*:533.

Allende, J. E., Mora, G., Gatica, M., and Allende, C. C. (1965): The role of metal ions in the formation of threonyl-soluble ribonucleic acid from threonyladenylate-enzyme complex. *Journal of Biological Chemistry, 240*:3229.

Appel, S. H., Davis, W., and Scott, S. (1967): Brain polysomes: Response to environmental stimulation. *Science, 157*:836.

Axelrod, A. E., and Trakatellis, A. C. (1964): Relationship of pyridoxine to immunological phenomena. *Vitamins and Hormones, 22*:591.

Bachelard, H. S., and Goldfarb, P. S. G. (1969): Adenine nucleotides and magnesium ions in relation to control of mammalian cerebral-cortex hexokinase. *Biochemical Journal, 112*:579.

Baxter, C. F. (1970): The nature of γ-aminobutyric acid. In: *Handbook of Neurochemistry,* Vol. 3. Edited by A. Lajtha. p. 289. Plenum Press, New York.

Baxter, C. F., and Tewari, S. (1970): Regulation by amino acids of protein synthesis in a cell-free system from immature rat brain: Stimulatory effect of γ-aminobutyric acid and glycine. In: *Protein Metabolism of the Nervous System,* edited by A. Lajtha. p. 439. Plenum Press, New York.

Beinert, H., Green, D. E., Hele, P., Hift, H., von Korff, R. W., and Ramakrishnan, C. F. (1963): The acetate activating enzyme system of heart muscle. *Journal of Biological Chemistry, 204*:35.

Berry, R. W. (1969): Ribonucleic acid metabolism of a single neuron correlation with electrical activity. *Science, 166*:1021.

Bloemendal, H., Littauer, U. Z., and Daniel, V. (1961): Transfer of soluble polynucleotides to microsomal RNA. *Biochimica et Biophysica Acta, 51*:66.

Brenner, M., DeLorenzo, F., and Ames, B. N. (1970): Energy change and protein synthesis. Control of aminoacyl transfer ribonucleic acid synthetases. *Journal of Biological Chemistry, 245*:450.

Campbell, M. K., Mahler, H. R., Moore, W. J., and Tewari, S. (1966): Protein synthesis systems from rat brain. *Biochemistry, 5*:1174.

Davis, S. D., Nelson, T., and Shepard, T. H. (1970): Teratogenicity of Vitamin B_6 deficiency: Omphalocele, skeletal and neural defects and splenic hypoplasia. *Science, 169*:1329.

Deutscher, M. P. (1968): Effect of polynucleotides on aminoacyl transfer ribonucleic acid synthetases. III. Inhibition of glutamyl transfer ribonucleic acid synthetase by natural polynucleotides. *Archives of Biochemistry, 125*:758.

Fleck, A., Shepherd, J., and Munro, H. N. (1965): Protein synthesis in rat liver: Influence of amino acids in diet on microsomes and polysomes. *Science, 150*:628.

Gershenovich, Z. S., Krichevskaya, A. A., and Shumskaya, V. I. (1965): Specificity of relations between γ-aminobutyric acid and brain proteins. *Doklady Akademii Nauk SSSR (Moskva), 162*:1415.

Grand, R. J., and Gross, P. R. (1970): Translation-level control of amylase and protein synthesis by epinephrine. *Proceedings of the National Academy of Sciences, 65*:1081.

Harbon, S., and Chapeville, F. (1970): Inhibition of protein synthesis in the reticulocytes by aminoacyl adenosine. *European Journal of Biochemistry, 13*:375.

Hoagland, M. B., Stephenson, M. L., Scott, J. F., Hecht, L. I., and Zamecnik, P. C. (1958): A soluble ribonucleic acid intermediate in protein synthesis. *Journal of Biological Chemistry, 231*:241.

Hydén, H., and Lange, P. W. (1970): Do specific biochemical correlates to learning process exist in brain cells? In: *Biochemistry of Simple Neuronal Models,* edited by E. Costa and E. Giacobini. p. 317. Raven Press, New York.

Lagerkvist, U., Rymo, L., and Waldenström, J. (1966): Structure and function of transfer ribonucleic acid. II. Enzyme-substrate complexes with valyl ribonucleic acid synthetase from yeast. *Journal of Biological Chemistry, 241*:5391.

Lebowitz, P., and Singer, M. (1970): Neurotrophic control of protein synthesis in the regenerating limb of the newt, *Triturus. Nature, 225*:824.

Lee, C. J. (1970): Biosynthesis and characteristics of brain protein and ribonucleic acid in mice subjected to neonatal infection or undernutrition. *Journal of Biological Chemistry, 245*:1998.

Loftfield, R. B., and Eigner, E. A. (1963): A radioactive hydroxamate method for determining rates of amino acid activation. *Biochimica et Biophysica Acta, 72*:372.

Luxoro, M. (1960): Incorporation of amino acids labeled with carbon-14 in nerve proteins during activity and recovery. *Nature, 188*:1119.

Mahler, H. R. (1970): In: *Protein Metabolism of the Nervous System.* Edited by A. Lajtha. p. 456, Plenum Press, New York.

Manchester, K. L. (1970): Assessment of effective ion concentration in systems synthesizing RNA and protein. *Biochimica et Biophysica Acta, 213*:532.

Metzger, H. P., Cuenod, M., Grynbaum, A., and Waelsch, H. (1966): Visual stimulation: Lack of effect on protein synthesis in monkey brain. *Life Sciences, 5*:1115.

Munro, H. N. (1964): General aspects of the regulation of protein metabolism by diet and by hormones. In: *Mammalian Protein Metabolism,* Vol. 1. Edited by H. N. Munro and J. B. Allison. p. 381. Academic Press, New York.

Munro, H. N. (1968): Regulation of liver ribosome function in relation to amino acid supply. In: *Regulatory Mechanisms for Protein Synthesis in Mammalian Cells,* edited by A. San Pietro, M. R. Lamborg, and F. T. Kenney. p. 183. Academic Press, New York.

Orrego, F., and Lipmann, F. (1967): Protein synthesis in brain slices. Effect of electrical stimulation and acidic amino acids. *Journal of Biological Chemistry, 242*:665.

Ritter, P. O., Kull, F. J., and Jacobsen, K. B. (1970): Aminoacylation of *Escherichia coli* valine transfer ribonucleic acid by *Neurospora crassa* phenylalanyl transfer ribonucleic acid synthetase in tris (hydroxymethyl) aminomethane hydrochloric acid and potassium cacodylate buffers. *Journal of Biological Chemistry, 245*:2114.

Rouget, P., and Chapeville, F. (1968): Reactions sequence of leucine activation catalysed by leucyl-RNA synthetase. *European Journal of Biochemistry, 4*:310.

Singh, U. B., and Talwar, G. P. (1967): Effect of the flicker frequency of light and other factors on the synthesis of proteins in the occipital cortex of monkey. *Journal of Neurochemistry, 14*:675.

Smith, D. W. E. (1969): The effect of salt solutions on the acceptance of amino acids by transfer ribonucleic acid. *Journal of Biological Chemistry, 244*:896.

Takeda, Y., and Igarashi, K. (1969): Polyamines and protein synthesis: IV. Stimulation of aminoacyl transfer RNA formation by polyamines. *Biochemical and Biophysical Research Communications, 37*:917.

Takeishi, K., Nishimura, S., and Ukita, T. (1967): Purification of histidine-specific transfer RNA from yeast. *Biochimica et Biophysica Acta, 145*:605.

Tewari, S., and Baxter, C. F. (1969): Stimulatory effect of γ-aminobutyric acid upon amino acid incorporation into protein by a ribosomal system from immature rat brain. *Journal of Neurochemistry, 16*:171.

Trakatellis, A. C., and Axelrod, A. E. (1964): Effect of pyridoxine deficiency upon valine incorporation into tissue proteins of the rat. *Journal of Nutrition, 82*:483.

Wong, J. T., and Mustard, M. (1969): Regulation of amino acid-specific transfer RNA's in *Escherichia coli*. *Biochimica et Biophysica Acta, 174*:513.

Wurtman, R. J., Shein, H. M., Axelrod, J., and Laren, F. (1969): Incorporation of ^{14}C tryptophan into ^{14}C protein by cultured rat pineals: Stimulation by l-norepinephrine. *Proceedings of the National Academy of Sciences, 62*:749.

Yang, S. S., and Sanadi, D. R. (1969): Changes in the distribution of transfer ribonucleic acid species specifically induced by thyroxine. *Journal of Biological Chemistry, 244*:5081.

Yu, C. T., and Hirsh, D. (1967): The stimulatory effect of ammonium or potassium ions on the activity of leucyl tRNA synthetase from *Escherichia coli*. *Biochimica et Biophysica Acta, 142*:149.

Zachmann, M., Tocci, P., and Nyhan, W. L. (1966): The occurrence of γ-aminobutyric acid in human tissues other than brain. *Journal of Biological Chemistry, 241*:1355–1358.

Zamenhof, S. van Marthens, E., and Margolis, F. L. (1968): DNA (cell number) and protein in neonatal brain: Alteration by maternal dietary protein restriction. *Science, 150*:322.

Advances in Biochemical Psychopharmacology, Vol. 4.
Raven Press, New York © 1972.

The Antivitamin B_6 Effect of Penicillamine: Clinical and Immunological Implications

Israeli A. Jaffe

*Department of Medicine,
New York Medical College,
New York, New York 10029*

Penicillamine (β-dimethyl cysteine) is now accepted as the treatment of choice for patients with Wilson's disease because of its copper-chelating properties (Walshe, 1960), and for cystinuric patients, in whom urinary cystine stone formation is prevented by the formation of a soluble penicillamine-cystine-mixed disulfide (Crawhall, Scowen, and Watts, 1963). It is currently under investigation in the treatment of rheumatoid arthritis, where it has been shown to produce clinical improvement associated with a fall in the sedimentation rate, a rise in hemoglobin, and marked reductions in the titer of rheumatoid factor in about 75% of patients with severe and unremitting disease (Jaffe, 1970; Zuckner, Ramsey, Dorner, and Gantner, 1970; Golding, Wilson, and Day, 1970).

I. ANTIVITAMIN B_6 EFFECT OF PENICILLAMINE

A. Biochemical Observations

Kuchinskas and du Vigneaud (1957) and Kuchinskas, Horvath, and du Vigneaud (1957) were the first to demonstrate growth retardation in the rat when L-penicillamine was added to the diet. This was completely reversible by the administration of pyridoxal (Wilson and du Vigneaud, 1950). The mechanism for the growth retardation was presumed to be the formation by penicillamine of a thiazolidine derivative with pyridoxal phosphate. Asatoor (1964) showed that the D-isomer also resulted in an antivitamin B_6 effect in the rat, and shortly thereafter we demonstrated that D-penicillamine was

TABLE 1. *The effect on the urinary excretion of the intermediary metabolites of tryptophan induced by 14 days of penicillamine administration*

		Control		D-Penicillamine (2 weeks)		Control		DL-Penicillamine (2 weeks)	
Patient	Condition	Xanthurenic acid	Kynurenine	Xanthurenic acid	Kynurenine	Xanthurenic acid	Kynurenine	Xanthurenic acid	Kynurenine
		(mg/24 hr)		(mg/24 hr)		(mg/24 hr)		(mg/24 hr)	
J. T.	normal	15	18	27	98				
P. E.	normal	9	15	56	49	15	13	243	164
B. L.	normal	20	26	88	75				
D. G.	rheumatoid arthritis	16	14	11	85	12	20	75	140
J. S.	rheumatoid arthritis	11	28	47	74	16	24	95	196
L. W.	rheumatoid arthritis	16	23	88	96				
R. B.	rheumatoid arthritis	20	60	60	117				
M. M.	rheumatoid arthritis	15	65	85	161				
C. P.	rheumatoid arthritis	14	17	97	73				

The data presented in this table compare the effect of D- and DL-penicillamine in a dosage of 2.0 g/day in nine study patients. Note that the values of Xa and Kyn are elevated by both drugs following tryptophan load. The greater antivitamin B_6 effect of DL-penicillamine is due to the presence of the 1-isomer in the racemic mixture. Kyn is the most sensitive of the metabolites.

capable of producing biochemical evidence of an antivitamin B_6 effect in man, in the dosage range commonly employed in clinical practice (Jaffe, Altman, and Merryman, 1964; Jaffe, 1969). We measured the 24 hr urinary excretion of xanthurenic acid (Xa) and kynurenine (Kyn) before and after a loading dose of 2 g of L-tryptophan. Patients receiving D-penicillamine showed an augmented excretion of these metabolites. If penicillamine was continued for prolonged periods, the biochemical defect persisted, and there was no tendency for spontaneous reversal of the phenomenon. When vitamin B_6 was added in adequate amounts, the metabolites promptly fell despite continuance of the drug (Fig. 1). Approximately 100 mg of vitamin B_6 was required to prevent completely the antagonism induced by 2.0 g of D-penicillamine per day. Without vitamin B_6 supplementation, more than 2 weeks was required for the return of Xa and Kyn to control values after discontinuing the drug. All patients received a normal hospital diet throughout these studies. Table 1 is a summary of the data obtained in nine subjects, three of whom were normal and six of whom had rheumatoid arthritis. This demonstration of the antivitamin B_6 effect of D-penicillamine in man has been subsequently confirmed (Gibbs and Walshe, 1969).

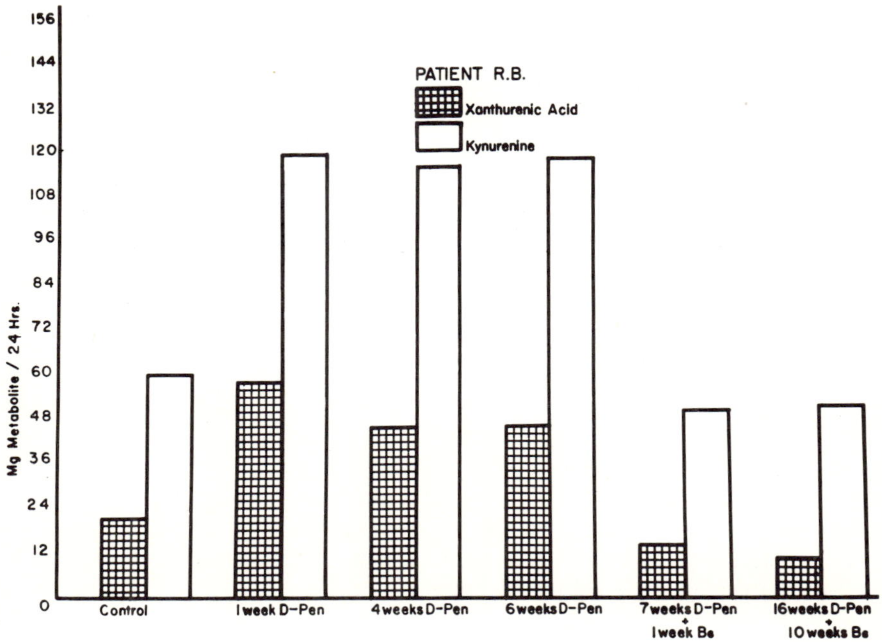

FIG. 1. This graph shows the augmented urinary excretion of Xa and Kyn induced by D-penicillamine at a dose of 2.0 g/day. After the sixth week, the penicillamine was continued and vitamin B_6 in a dose of 100 mg/day was added. This resulted in a fall of Xa and Kyn to control values.

B. Clinical Manifestations

The clinical manifestations of the biochemical antagonism have been minimal. Several patients with Wilson's disease were reported to have developed epileptic seizures while undergoing penicillamine therapy, at a time when their other neurological symptoms resulting from the Wilson's disease were clearing (Gibbs and Walshe, 1969; Smith and Gallagher, 1970). In these situations, the concomitant administration of vitamin B_6 resulted in a prompt normalization of the EEG pattern and disappearance of the seizures despite continuation of the penicillamine. It has also been shown that in an infant born with presumed pyridoxine-dependent epileptic seizures, the administration of 140 mg of D-penicillamine resulted in a grand mal seizure which was immediately terminated after the intravenous administration of 25 mg of vitamin B_6. It was suggested that penicillamine administration might provide a useful provocative test in infants who have refractory epilepsy, when the diagnosis of pyridoxine dependent seizures is being considered (Swaiman and Milstein, 1970).

Two patients on long-term penicillamine therapy for severe rheumatoid arthritis, whose nutritional status was marginal, have been observed by the author to develop "stocking glove" hypasthesia which was reversed by vitamin B_6 despite continuance of the penicillamine. One patient with Wilson's disease who received DL-penicillamine developed optic axial neuritis, reversed by the administration of vitamin B_6 (Tu, Blackwell, Cooper, and Chen, 1964).

It had been suggested that the nephropathy induced by penicillamine, which is the most serious toxicity of the drug, was encountered only in patients given the DL form (Sternlieb, 1966). Subsequent reports, however, have shown that proteinuria and nephrotic syndrome also occur after treatment with the D-isomer even when vitamin B_6 supplements are given (Jaffe, Treser, Suzuki, and Ehrenreich, 1968), suggesting that vitamin B_6 deficiency is not responsible for the renal lesion.

None of the other toxic and untoward effects of penicillamine therapy, such as leukopenia, skin rash, drug fever, and diminution in taste acuity, has been favorably influenced by concomitant vitamin B_6 administration. Thus it may be seen that *clinical* manifestations of vitamin B_6 deficiency induced by the drug have been few when considering the large number of patients to whom penicillamine has been given. It is, therefore, our routine practice to administer vitamin B_6 supplements to *adult* patients on long-term penicillamine therapy only if their nutritional status is sub-optimal, or if a daily dose of more than 2.0 g of the drug is required.

II. VITAMIN B_6 DEFICIENCY AND IMMUNOSUPPRESSION

The suppressive effects of vitamin B_6 deficiency on the immune apparatus in experimental animals have been well documented, and their possible relevance to patients treated with the drug will be considered. In these animal studies, it has been shown that vitamin B_6 deficiency induced either by dietary means or vitamin B_6 analogues, or by both simultaneously, results in a marked inhibition of circulating antibody titer and delayed hypersensitivity response following antigen administration. This subject has been exhaustively studied and reviewed by Axelrod and Trakatellis (1964). These investigators concluded that pyridoxine deficiency both *in vivo* and *in vitro* results in an inhibition of incorporation of amino acids into tissue protein with a decreased biosynthesis of DNA, RNA, and particularly messenger RNA. They believe that this inhibition of nucleic acid biosynthesis is probably the mechanism by which vitamin B_6 deficiency interferes with cellular proliferation and synthesis of specific immunoglobulins. Although there are data from studies in animals that show an impairment in circulating antibody response following *penicillamine* administration, the role of vitamin B_6 antagonism in these investigations was not clearly defined (Hubner and Gengozian, 1965). Unfortunately, attempts to apply analogue-antagonists of vitamin B_6 to the clinical problems of "autoimmune" disease and tissue homograft rejection have not been successful (Rosen, Mihich, and Nichol, 1964). One explanation offered for this failure has been that the agents presently available are relatively weak inhibitors, whose effects are readily reversible in the presence of adequate amounts of the vitamin in the diet. It has also been suggested that were a truly potent antagonist or inhibitor of vitamin B_6 to be given to man, the side effects—particularly upon the central nervous system—might preclude its application clinically (Rosen *et al.*, 1964). Some of these problems were encountered by Gellhorn and Jones (1949) in clinical cancer chemotherapy.

III. SIGNIFICANCE OF PENICILLAMINE-INDUCED VITAMIN B_6 DEFICIENCY IN RHEUMATOID ARTHRITIS PATIENTS.

The experimental evidence in animals of the inhibitory effect of vitamin B_6 antagonists on immune responses was of particular interest in view of the previously mentioned effect of long-term D-penicillamine treatment in adult patients with peripheral rheumatoid arthritis. Rheumatoid disease, at least in part, appears to be caused by immune mechanisms. Approximately 80% of such patients have, in their serum, an antibody called the rheumatoid factor (RF) which is detected by the latex fixation test and which is a 19s gamma macroglobulin that reacts *in vitro* and *in vivo* with 7s gamma globulin. The rheumatoid

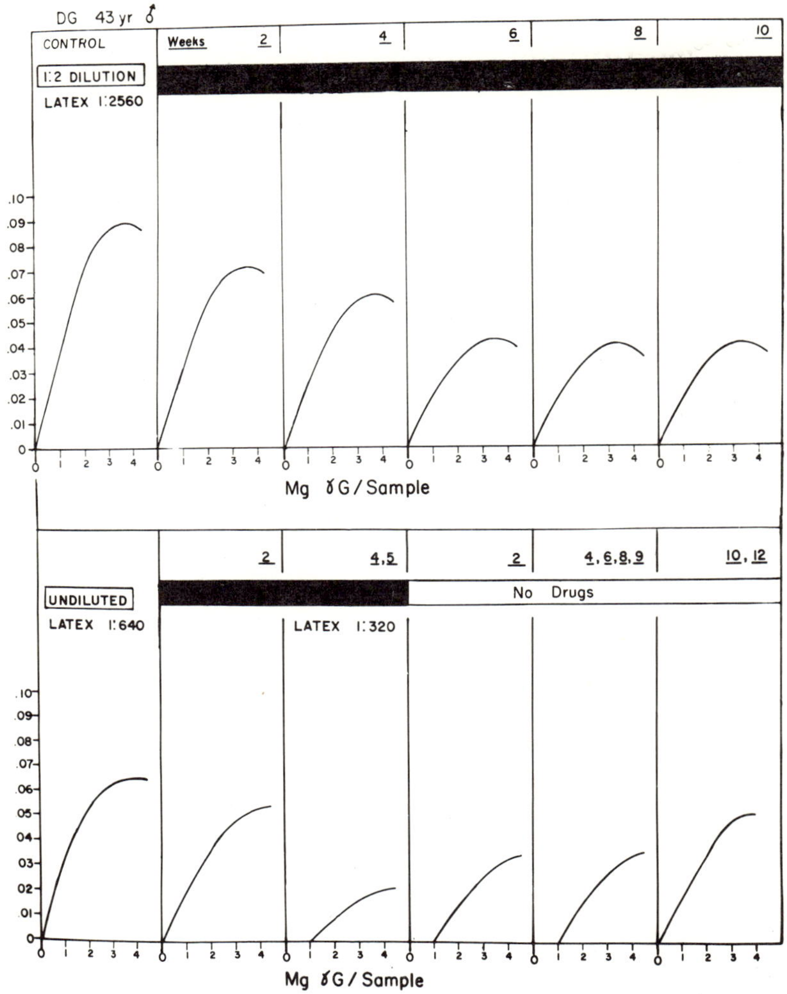

FIG. 2. These curves represent changes in the level of rheumatoid factor in the serum from an arthritis patient given penicillamine at a dosage of 2 g/day (solid line). The height of the precipitin curve and the degree of serum dilution relate to the amount of RF present. Note the gradual fall in titer which persisted for 2–1/2 months after the drug was stopped.

factors and their biological significance have been the subject of a recent review (Bartfeld and Epstein, 1969). Rheumatoid patients who have high titers of circulating RF tend to have the poorest prognosis, experience fewer spontaneous remissions, and are most likely to have mutilating disease. D-Penicillamine is believed to be more effective than any other pharmacological agent in producing reductions in the titer of RF following prolonged therapy (Jaffe, 1970), even when it is compared to such cytotoxic "immunosuppressant" drugs as cyclophosphamide (Mainland, Decker, Healey, Mikkelsen, Wallace, and Zvaifler, 1970).

RF titers fall very slowly after the institution of penicillamine therapy, and the reduced titers tend to persist for many months after the drug has been discontinued (Fig. 2). This indicates that the site of drug action is on the cellular level rather than on pre-formed macroglobulin. Since RF is an antibody, and D-penicillamine has antivitamin B_6 properties, the hypothesis was entertained that the mechanism of action of penicillamine with respect to RF, and perhaps to the clinical improvement obtained in many of these patients, was due to immunosuppression by the mechanism of penicillamine-induced vitamin B_6 antagonism. Three sets of experiments were performed to test this hypothesis and the results were negative (Jaffe, 1965). In these studies, the tryptophan load test and the excretion of Kyn and Xa were used as the indicators of vitamin B_6 deficiency. One patient was given D-penicillamine supplemented with 100 mg of pyridoxine daily. The urinary metabolites after tryptophan load remained normal, suggesting that the vitamin B_6 antagonism had been corrected, but the patient experienced a marked fall in RF titer precisely as had been previously found when the vitamin was not administered as part of the penicillamine regimen (Jaffe, 1965). We have subsequently confirmed in many patients that the administration of 100 mg of vitamin B_6 daily with penicillamine does not prevent the clinical improvement in the arthritis nor the drop in RF titer (Fig. 3). A second patient was given L-penicillamine in a dose of 200 mg/day. The excretion of Xa and Kyn was the same as that produced by 2,000 mg/day of the D-isomer, indicating a comparable degree of antivitamin B_6 effect; however, there was no clinical benefit or fall in RF titer. Finally, a third rheumatoid patient was given isoniazid in a dosage of 1,200 mg/day, which elevated Xa and Kyn to approximately the same range as that produced by 2,000 mg/day of D-penicillamine. The patient failed to sustain clinical or laboratory improvement during his course of isoniazid therapy (Jaffe, 1965).

It would certainly appear from the data derived in these three sets of experiments that the mechanism for the suppression of the clinical and immunological findings in rheumatoid patients by chronic penicillamine therapy is not drug-induced vitamin B_6 antagonism. Patients with Wilson's disease and cystinuria, who have taken penicillamine for many years, do not behave as though they have been "immunosuppressed." They do not have increased susceptibility to infection nor do they develop the opportunistic infections so frequently found in patients who are on prolonged antimetabolite or cytotoxic therapy (Schwartz, 1969).

Yet, the previously cited clinical studies and observations may not have fully resolved the question. Is it not possible that *selective* vitamin B_6 antagonism could be induced by penicillamine at the level of the hyperactive immunocyte which is undergoing sustained antigenic stimulation and hence a markedly increased vitamin B_6 requirement? Indeed, Coursin (1964) has

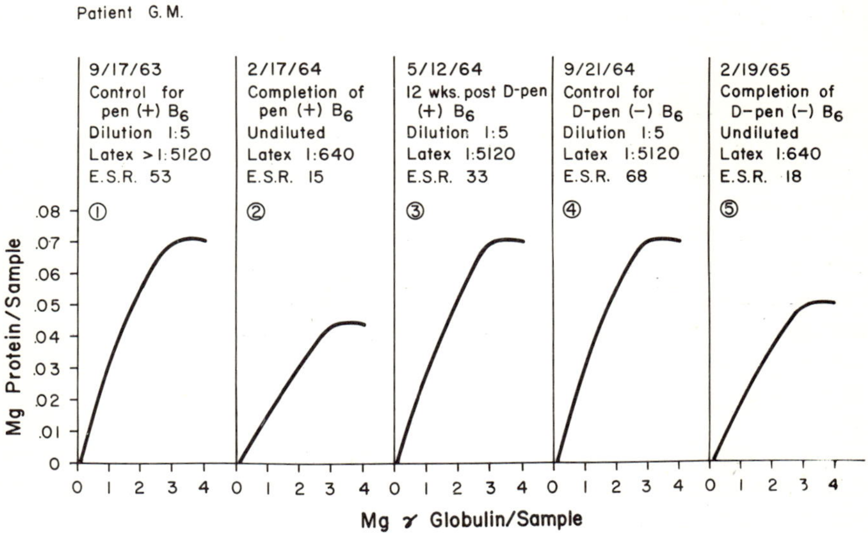

FIG. 3. This patient was given two courses of penicillamine. The first course from 9/17/63 to 2/17/64 was given with vitamin B_6. The second course from 9/21/64 to 2/19/65 consisted of penicillamine alone without pyridoxine supplementation. Note that the fall in precipitin curves, latex titer, and erythrocyte sedimentation rate was approximately the same during both courses of treatment, indicating that vitamin B_6 supplementation did not prevent the response induced by penicillamine.

suggested that the "finding of a negative tryptophan load test does not preclude the possibility of abnormalities of *other* vitamin B_6-dependent enzyme systems that may have *higher* requirements for their operation than are necessary for the degradation of tryptophan."

The absence of an increased incidence of infection in penicillamine-treated patients could be explained by the fact that leukopenia and impaired leukocyte response are not features of penicillamine therapy, as they are with conventional antimetabolite and cytotoxic immunosuppressive agents (Schwartz, 1969). When leukopenia is infrequently encountered during penicillamine therapy, it is idiosyncratic rather than dose related and often disappears despite continuance of the drug (Jaffe, 1965).

Thus, all long-term penicillamine-treated patients may be "weakly immunosuppressed" with regard to their ability to respond to antigenic challenge, but they retain normal leukocyte capability, and so this does not result in an impairment of resistance to infectious agents. Perhaps the antibody response (the rheumatoid factor) to a low-grade, sustained antigenic stimulus, such as that which is presumed to exist in rheumatoid arthritis patients, could be inhibited by penicillamine (which is a vitamin B_6 antagonist in man in the presence of a regular diet). As indicated, this might well occur despite normalization of the tryptophan degradation pathway with supplements of vitamin B_6, on the presumption that the coenzyme requirements for the vitamin by the antigenically stimulated immunocyte are still greater than that of kynureninase and the other vitamin B_6-dependent enzymes required for the catabolism of tryptophan. Since we do not have a comparable means for measuring the amount of vitamin B_6 needed by the antigenically stimulated, immunologically competent cell, proof for such a scheme is necessarily lacking. Although our means for testing the integrity of the immune apparatus in man are still crude, detailed studies of the humoral and cellular antibody responses of long-term penicillamine-treated patients would appear warranted in an attempt to provide indirect support for such an hypothesis.

IV. SUMMARY

The D-isomer of penicillamine has been shown to be a biochemical antagonist to vitamin B_6 in man in the dosage range in which the drug is clinically employed. The side effects of penicillamine which relate to the central and peripheral nervous system are due to drug-induced vitamin B_6 deficiency, but they occur only infrequently and are readily reversed by supplementation with vitamin B_6 despite continuance of the drug. It is suggested that a vitamin B_6 deficiency state may lead to a moderate degree of immunosuppression in patients on long-term penicillamine therapy, as it does in some animal models. This does not manifest itself clinically by increased susceptibility to infection, probably because the degree of inhibition is weak and the leukocyte response is intact. The hypothesis is offered that the sustained fall in rheumatoid factor titer in arthritis patients treated with penicillamine could be due to a suppressive effect of the drug on antibody synthesis via the mechanism of vitamin B_6 antagonism, even when the tryptophan-kynurenine pathway has been corrected with supplements of the vitamin.

Acknowledgment. This work was supported by the U.S. Army Medical Research and Development Command grant number DA-49-193MD-2445.

REFERENCES

Asatoor, A. M. (1964): Pyridoxine deficiency in the rat produced by D-penicillamine. *Nature, 203*:1382.

Axelrod, A. E., and Trakatellis, A. C. (1964): Relationship of pyridoxine to immunological phenomena. In: *Vitamins and Hormones,* vol. 22. Edited by R. S. Harris, I. G. Wool, and J. A. Loraine, p. 591. Academic Press, New York.

Bartfeld, T. T., and Epstein, W. V. (1969): Rheumatoid factors and their biological significance. *Annals of the New York Academy of Sciences, 168*:1.

Coursin, D. B. (1964): Vitamin B_6 metabolism in infants and children. In: *Vitamins and Hormones,* vol. 2. Edited by R. S. Harris, I. G. Wool, and J. A. Loraine. p. 755. Academic Press, New York.

Crawhall, J. C., Scowen, E. F., and Watts, R. W. E. (1963): Effect of penicillamine on cystinuria. *British Medical Journal, 1*:588.

Gellhorn, A., and Jones, L. O. (1949): Pyridoxine deficient diet and desoxypyridoxine in the therapy of lymphosarcoma and acute leukemia in man. *Blood, 4*:60.

Gibbs, K., and Walshe, J. M. (1969): Interruption of the tryptophan-nicotinic acid pathway of penicillamine-induced pyridoxine deficiency in patients with Wilson's disease and in experimental animals. *Annals of the New York Academy of Sciences, 166*:158.

Golding, J. R., Wilson, J. V., and Day, A. T. (1970): Observation on the treatment of rheumatoid disease with penicillamine. *Postgraduate Medical Journal, 46*:599.

Hübner, K. F., and Gengozian, N. (1965): Depression of the primary immune response by dl-penicillamine. *Proceedings of the Society for Experimental Biology and Medicine, 118*:561.

Jaffe, I. A., Altman, K., and Merryman, P. (1964): The antipyridoxine effect of penicillamine in man. *Journal of Clinical Investigation, 43*:1869.

Jaffe, I. A. (1965): The effect of penicillamine on the laboratory parameters in rheumatoid arthritis. *Arthritis and Rheumatism, 8*:1064.

Jaffe, I. A., Treser, G., Suzuki, Y., and Ehrenreich, T. (1968): Nephropathy induced by D-penicillamine. *Annals of Internal Medicine, 69*:549.

Jaffe, I. A. (1969): Antivitamin B_6 effect of D-penicillamine. *Annals of the New York Academy of Sciences, 166*:57.

Jaffe, I. A. (1970): The treatment of rheumatoid arthritis and necrotizing vasculitis with penicillamine. *Arthritis and Rheumatism, 13*:436.

Kuchinskas, E. J., and du Vigneaud, V. (1957): An increased vitamin B_6 requirement in the rat on a diet containing L-penicillamine. *Archives of Biochemistry and Biophysics, 66*:1.

Kuchinskas, E. J., Horvath, A., and du Vigneaud, V. (1957): An anti-vitamin B_6 action of L-penicillamine. *Archives of Biochemistry and Biophysics, 68*:69.

Mainland, D., Decker, J. L., Healey, L. A., Mikkelsen, W. M., Wallace, S. L., and Zvaifler, N. J. (1970): A controlled trial of cyclophosphamide in rheumatoid arthritis. *The New England Journal of Medicine, 283*:883.

Rosen, F., Mihich, E., and Nichol, C. A. (1964): Selective metabolic and chemotherapeutic effects of vitamin B_6 antimetabolites. In: *Vitamins and Hormones,* vol. 22. Edited by R. S. Harris, I. G. Wool, and J. A. Loraine. p. 609. Academic Press, New York.

Schwartz, R. S. (1969): Therapeutic strategy in clinical immunology. *The New England Journal of Medicine, 280*:367.

Smith, D. B., and Gallagher, B. B. (1970): The effect of penicillamine on seizure threshold. *Archives of Neurology, 23*:59.

Sternlieb, I. (1966): Penicillamine and the nephrotic syndrome. *The Journal of the American Medical Association, 198*:1311.

Swaiman, K. F., and Milstein, J. M. (1970): Pyridoxine dependency and penicillamine. *Neurology, 20*:78.

Tu, J. B., Blackwell, R. Q., Cooper, W. C., and Chen, Y. H. (1964): Studies of pyridoxal-penicillamine antagonism in the human. *Biochemical Pharmacology, 13*:1527.

Walshe, J. M. (1960): Treatment of Wilson's disease with penicillamine. *The Lancet, 1*:118.

Wilson, J. E., and du Vigneaud, V. (1950): Inhibition of the growth of the rat by L-penicillamine and its prevention by aminoethanol and related compounds. *Journal of Biological Chemistry, 184*:63.

Zuckner, J., Ramsey, R. H., Dorner, R. W., and Gantner, G. E., Jr. (1970): D-penicillamine in rheumatoid arthritis. *Arthritis and Rheumatism, 13*:131.

Advances in Biochemical Psychopharmacology, Vol. 4.
Raven Press, New York © 1972.

Platelet Transaminase

David J. Boullin

*Laboratory of Preclinical Pharmacology,
National Institute of Mental Health,
Saint Elizabeths Hospital,
Washington, D.C. 20032*

I. PLATELET TRANSAMINASE

It is impossible to study many aspects of brain function in man because tissue samples are so difficult to obtain. When specific lesions are noted in brain disease, we often try to find similar changes in other, more readily accessible tissues. It would be extremely helpful if changes occurring in brain tissues could be monitored, even in part, by studies with peripheral organs. There is a rationale for such a proposal because many cells have certain basic properties in common, including, for example, membrane structure and mechanisms of ion transport. In agreement with this, Barbeau (1969) has advanced the view that in Parkinson's disease the defects in dopamine binding and metabolism are not confined to the brain. This certainly seems to be true for mechanisms for uptake and binding of dopamine by cells, because we have found that there is a binding defect for dopamine in blood platelets from parkinsonian subjects. In dopamine-loaded platelets from parkinsonian subjects, the rate of loss of amine is two to seven times greater than normal (Boullin and O'Brien, 1970*a*). Although the precise biochemical lesion in the strategic nigro-striatal pathway in Parkinson's disease is unknown, a plausible explanation is that there is a binding abnormality in dopaminergic neurons. If this is true and if such an abnormality is also found in the blood platelets, then the validity of the platelet model is established in regard to binding of dopamine in neurons. In addition, abnormalities in 5-hydroxytryptamine uptake and binding have been found in platelets from children with Down's syndrome, where disorders in serotonin (5-HT) storage may also occur in brain neurons (Boullin and O'Brien, 1970*b*). Thus the use of the platelet as a model for brain neuron transmitters may have a wide application.

Until now the concept of the platelet as a model has been confined to the uptake, binding, and metabolism of those amines which presumably function as neurohumoral transmitters in the brain. Can the use of the platelet model be extended into the realm of enzymology? As a starting point, I proposed to study platelet transaminase, because there is currently considerable interest in brain transaminase. In particular, the recent work of Axelrod and his colleagues (Black and Axelrod, 1968, 1969), showing that noradrenaline can inhibit tyrosine transaminase and that noradrenaline depletion can induce the enzyme, may indicate that transamination reactions may have some hitherto undisclosed role, regulating various aspects of neurohumoral transmitter function. Also it is well known from the work of La Due, Wroblewski, and Karmen (1954), Karmen, Wroblewski, and La Due (1955), and Koj, Zgliczynski, and Frendo (1960) that the activities of the transaminase enzymes in platelets and other blood cells are increased in myocardial infarction and may serve as a diagnostic aid.

The transamination of certain amino acids may be of particular interest to the neurochemist, because these substances serve as precursors of neurohumoral transmitter amines. Such amino acid precursors include tyrosine, phenylalanine, tryptophan, and 5-hydroxytryptophan, while γ-amino butyric acid may act directly as chemical transmitter substance. If the platelet were to show active transamination of various α-amino acids and the enzymes possessed similar biochemical characteristics to the brain transaminases, then changes in platelet transaminases might serve as important indicators of enzymatic changes within the brain itself.

II. METHODS

Human blood was collected into plastic centrifuge tubes containing either disodium edetate or sodium citrate to give final concentrations of 30 and 12.9 μmoles/ml, respectively. Platelet-rich plasma was obtained by centrifugation (Boullin and O'Brien, 1969), and the number of cells/ml counted with a Model B Coulter Counter (Coulter Electronics, Inc., Hiahlea, Fla.). The platelets were then separated from the plasma by high-speed centrifugation (Boullin and O'Brien, 1969) The supernatant plasma was decanted and traces of the solution removed from the centrifuge tubes. The platelets were resuspended in 150 μmoles/ml KH_2PO_4, triethanolamine buffer, or Krebs solution having the following composition (μmoles/ml): NaCl, 121; KCl, 4.9; K_2HPO_4, 1.2; $MgSO_4$, 1.2; $NaHCO_3$, 25.5; and glucose, 11.0 (calcium was omitted). The pH of all buffers was 7.4. In some experiments the platelets were disrupted by sonification using a Biosonic II Sonifier fitted with a microtip. A standard setting of 50 was used with periods of short-duration sonification, totally not more than 15 sec.

A. Transaminase

The radiochemical assay procedure used is based on the principle that radioactively labeled α-amino acids will be transaminated to labeled α-keto acids which can then be separated from the parent α-amino acid by ion exchange chromatography. A similar procedure has previously been used by Hall and Kravitz (1967) to study γ-amino butyric acid transaminase in the lobster nervous system and by Bartholini, Gey, and Pletscher (1970) to study liver tyrosine transaminase in the rat.

Basic procedure: A 2 ml portion of the incubation medium contained the following: intact or disrupted platelets derived from 2 ml platelet-rich plasma; 150 μmoles/ml KH_2PO_4 or triethanolamine, or 154 μmoles/ml Krebs solution, as stated in Results; and 200 nmoles/ml α-ketoglutaric acid or oxaloacetic acid. In some experiments 200 nmoles/ml pyridoxal 5′-phosphate (PLP) was also included. The transamination reaction was observed using labeled amino acids with specific activities indicated: L-aspartic acid-^{14}C (U), 229 mc/mmole; L-leucine-^{14}C (U), 344 mc-mmole; L-glutamic acid-^{14}C (U), 260 mc/mmole; γ-amino butyric acid-1-^{14}C, 1.64 mc/mmole; DL-5-hydroxytryptophan-3-^{14}C, 21.8 mc/mmole; L-tyrosine^{14}C(U), 260 mc/mmole; L-phenylalanine-3-^{14}C, 25 mc/mmole, All were supplied by Amersham/Searle, Inc. (Arlington Heights, Ill.), except for leucine and γ-amino butyric acid which were from New England Nuclear Corp. (Boston, Mass.). In most experiments the concentration of α-amino acid was 1 nmole/ml. Actual concentrations are given in the text. All incubations were in air at 37°C in a metabolic shaker, and usually the incubation time was 60 min. Control incubations were made with either boiled platelets or tissue-free solutions. At the end of incubation, the reaction was stopped by addition of 2 ml of 20% trichloroacetic acid. The incubation tubes were centrifuged at 20,000 g for 5 min to precipitate protein. Then 3 ml of the supernatant solution was applied to 23 × 26 mm columns of Dowex 50-X-8 (200–400 mesh, NH_4^+ form) cation-exchange resin. The columns were washed with 3 ml distilled water. Samples of the acidified incubation medium (containing both unmetabolized amino acid and transaminated products), effluent, and wash fractions (containing only the transaminated products) were counted in a Triton-X-100-based phosphor in a liquid scintillation spectrometer.

III. RESULTS

The validity of the radiochemical procedure as a transaminase assay was established using authentic glutamic oxaloacetic transaminase from pig heart (Sigma Chemical Co., St. Louis, Mo.). Addition of 0.1 to 10 μg of this enzyme to a standard incubation mixture containing 10 nmoles/ml L-glutamic acid-^{14}C and PLP resulted in quantitative recovery of radioactivity in the effluent and wash fractions; with 5 and 10 μg there was 100% recovery in 60 min (Fig. 1), indicating complete conversion of glutamic-^{14}C acid to α-keto glutarate-^{14}C.

In the first experiments with platelets, I studied the effects of various incubation media on transamination of L-glutamic acid. Disrupted cells were incubated with 150 μmoles/ml KH_2PO_4 or triethanolamine or 154 μmoles/ml Krebs solution, pH 7.4. All solutions were calcium-free. Figure 2 shows that maximal transamination of L-aspartic acid occurred with KH_2PO_4 buffer, and therefore all subsequent experiments were made with this medium. Next, I utilized the enzyme glutamic oxaloacetic transaminase having the general reaction:

$$\text{Aspartate} + \alpha\text{-ketoglutarate} \gtrless \text{glutamate} + \text{oxaloacetate} \quad (1)$$

The reaction was run in both directions using either labeled L-glutamic acid or L-aspartic acid as amino-donor and oxaloacetic acid or α-ketoglutaric acid as amino acceptors. Table 1 shows that little transaminase activity was detectable with intact platelets, although addition of PLP did produce an

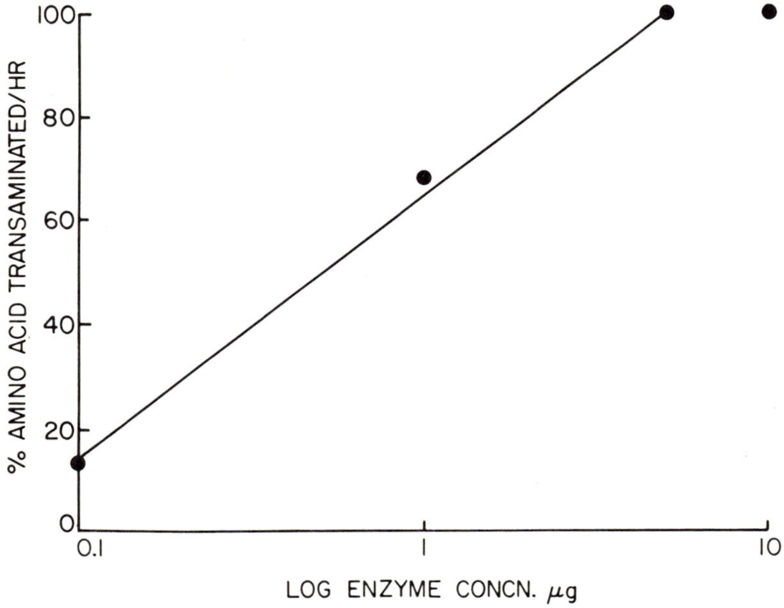

FIG. 1. Radiochemical determination of transamination of L-glutamic acid-^{14}C using glutamic oxaloacetic transaminase (GOT) from pig heart. The incubation media contained 150 μmols/ml KH_2PO_4 pH 7.4, 200 nmols/ml oxaloacetic acid, 10 nmols/ml L-glutamic acid ^{14}C, 200 nmols/ml PLP, and 0.1 to 10 μg enzyme (GOT). Incubations were for 60 min as described in Methods. Results are plotted as % amino acid transaminated/hr as ordinate, against log enzyme concentration (μg) abscissa, and are the mean of 2 determinations.

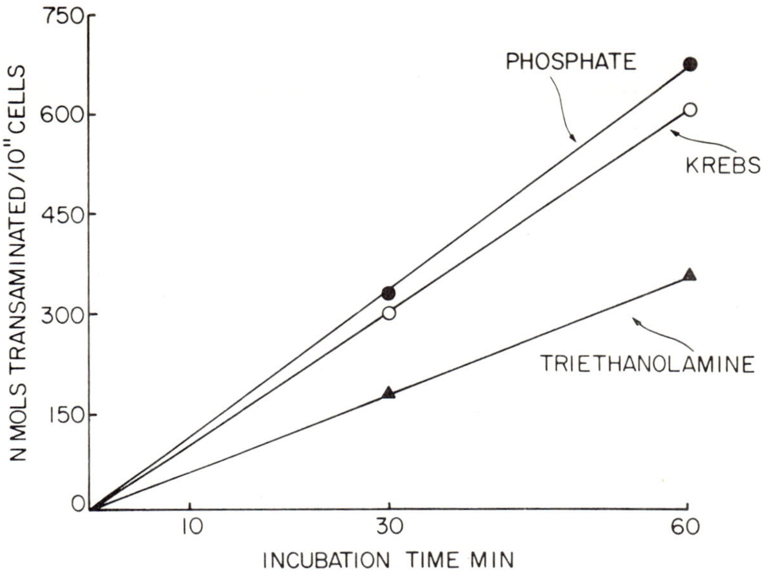

FIG. 2. Effect of incubation medium on transamination of L-aspartic acid. Sonified platelets were incubated in 150 μmols/ml KH_2PO_4 (●), triethanolamine (▲), or 154 μmols/ml Krebs solution (o) in the presence of 200 nmols/ml α-oxaloacetic acid and 200 nmols/ml PLP for 30 or 60 minutes at 37°C as described in Methods. Values were obtained in 2 experiments.

TABLE 1. *Transamination of* L-*glutamic and* L-*aspartic acids by normal human platelets (nmoles transminated products formed/10^{11} cells/hr)*

α-Amino acid	α-keto acid	Intact Cells		Disrupted Cells	
		+ PLP	− PLP	+ PLP	− PLP
glutamic	oxaloacetic	ND	ND	67 ± 5	59 ± 8
glutamic	ketoglutaric	7.9 ± 0.8	3.5 ± 0.3	92 ± 8	63 ± 10
aspartic	ketoglutaric	ND	ND	603	497

Platelets were incubated in 150 μmoles/ml phosphate buffer for 60 min as described in Methods. The concentrations of reagents were: α-amino acids, 1 nmole/ml; α-ketoacids and pyridoxal 5-phosphate (PLP), 200 nmoles/ml. Values are the mean ± S.E. of four determinations or mean of two determinations. ND = no detectable activity.

increase in the transamination of glutamic acid, with α-ketoglutaric acid as amino acceptor. In the experiments with disrupted platelets, transamination was most rapid when the reaction shown in equation 1 was run from left to right, involving conversion of labeled aspartic acid to oxaloacetic acid. All reactions were stimulated by PLP and were completely inhibited by the transaminase inhibitors amino oxyacetic acid and hydroxylamine (1 μmole/ml).

Using this pyridoxal phosphate-dependent reaction, I thought it worthwhile to see if any abnormalities existed in the transamination of glutamic acid and leucine in patients diagnosed as chronic schizophrenics. Leucine was also studied because we have found some transamination of this amino acid by a pyridoxal phosphate-dependent platelet enzyme (D. J. Boullin and M. Votavova, *unpublished observations*, and this paper). Although there is no evidence for disorders in cerebral transaminase in mental disease, it is known that various serum enzymes are increased in schizophrenic and other psychotic patients (Meltzer, 1968). Table 2 shows that platelet transamination of aspartic and glutamic acids and leucine is unchanged in schizophrenic subjects whether experiments are made with whole cells or with disrupted cells with or without added PLP. Clearly, therefore elevation of enzymes, which includes those released from platelets, is not a generalized phenomenon in schizophrenia.

TABLE 2. *Transaminase in whole and disrupted platelets (nmoles transaminated products formed/10^{11} cells/hr)*

Amino acid	Normal		Schizophrenic	
	+ PLP	− PLP	+ PLP	− PLP
a. whole platelets				
glutamic	43.7 ± 11.4	37.2 ± 9.5	58.5 ± 9.6	38.2 ± 8.8
leucine	24.3 ± 1.6	22.6 ± 1.2	22.3 ± 1.8	19.3 ± 0.9
b. disrupted platelets				
glutamic	90 ± 7	84 ± 7	95 ± 9	89 ± 8
leucine	32.6 ± 2.2	19.6 ± 2.1	29.4 ± 2.7	17.8 ± 1.8

Intact or disrupted platelets were incubated with 1 nmole/ml ^{14}C-glutamic acid or ^{14}C-leucine, 200 nmoles/ml α-ketoglutaric acid, ± 200 nmoles/ml PLP as described in Methods. Results are the mean ± S.E. of the mean of eight experiments with normal subjects and four experiments with schizophrenic subjects.

Because other tissues are known to transaminate many amino acids, it seemed interesting to find out if such reactions also occurred in blood platelets. Therefore, transamination of the following amino acids was investigated: leucine, tyrosine, 5-hydroxytryptophan, phenylalanine, and γ-aminobutyric acid. Only leucine and tyrosine were transaminated by intact platelets incubated with 1 nmole/ml of the amino acid for 1 hr (Table 3). However, some activity was demonstrable with all the amino acids when transamination was studied in disrupted cells. Activity was highest with leucine and lowest with γ-aminobutyric acid (Table 4). Recently Semba and Civen (1970) found that tyrosine and 5-hydroxytryptophan transaminases showed comparable activity in rat brain. On the basis of my experiments, I considered that the reactions which could be most profitably investigated further involved transamination

TABLE 3. *Transamination of amino acids by whole platelets (nmoles transaminated products formed/10^{11} cells/hr)*

Amino acid	+ PLP	− PLP
leucine (n = 8)	24.3 ± 1.6	22.6 ± 1.2
tyrosine (n = 4)	9.5 ± 1.2	8.4 ± 0.8

Platelets were incubated with 1 nmole/ml amino acid, 150 μmoles/ml KH_2PO_4 buffer, pH 7.4, 200 nmoles/ml α-ketoglutaric acid, ± 200 nmoles/ml PLP for 60 min as described in Methods. Results are the mean ± S.E. of the number of experiments (n) indicated. No detectable transamination was noted with the following amino acids: phenylalanine, 5-hydroxytryptophan, or γ-amino butyric acid.

of aspartic and glutamic acids and leucine. Figure 3 shows the time course of transamination of these substances. Enzymatic activity is linear for 60 min, and the rate of transamination of aspartic acid is approximately 19 times faster than leucine.

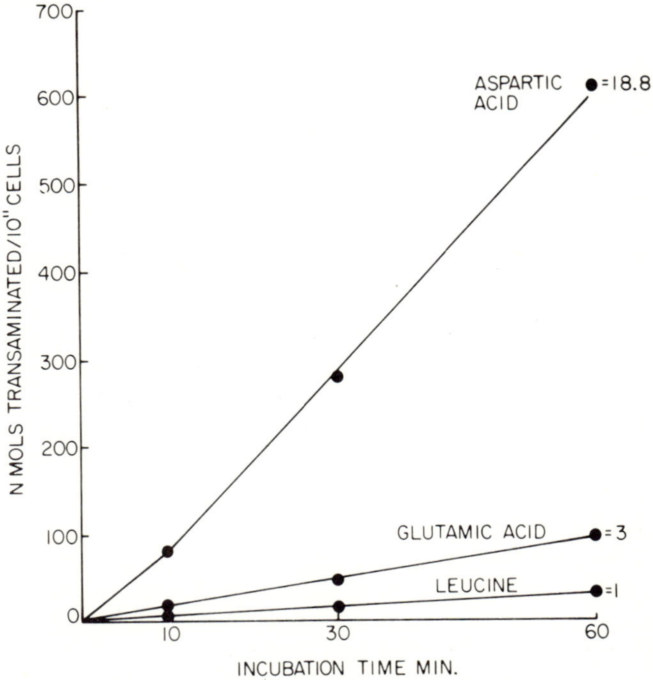

FIG. 3. Transamination of aspartic acid, glutamic acids, and leucine by sonified platelets. Cells were incubated with 1 nmol/ml of ^{14}C-labeled amino acids (glutamic acid, aspartic acid, or leucine), 200 nmols/ml of oxaloacetate, and 200 nmols/ml ml of PLP for 30–60 minutes as described in Methods. Values are the mean from two experiments.

TABLE 4. *Transamination of amino acids by disrupted platelets (nmoles transaminated products formed/10[11] cells/hr)*

Amino acid	+ PLP	− PLP
leucine	32.6 ± 2.2	19.6 ± 2.1
DL 5-hydroxytryptophan	29.9	27.9
tyrosine	16.1	15.3
phenylalanine	4.8	5.0
γ-amino butyric acid	1.1	1.3

Sonified platelets were incubated in 150 μmoles/ml phosphate buffer, pH 7.4, containing 200 nmoles/ml α-ketoglutaric acid, and 1 nmole/ml of one of the ^{14}C-labeled amino acids shown in the Table. The concentration of PLP was 200 nmoles/ml. Values are the mean ± S.E. of four determinations or the mean of two determinations.

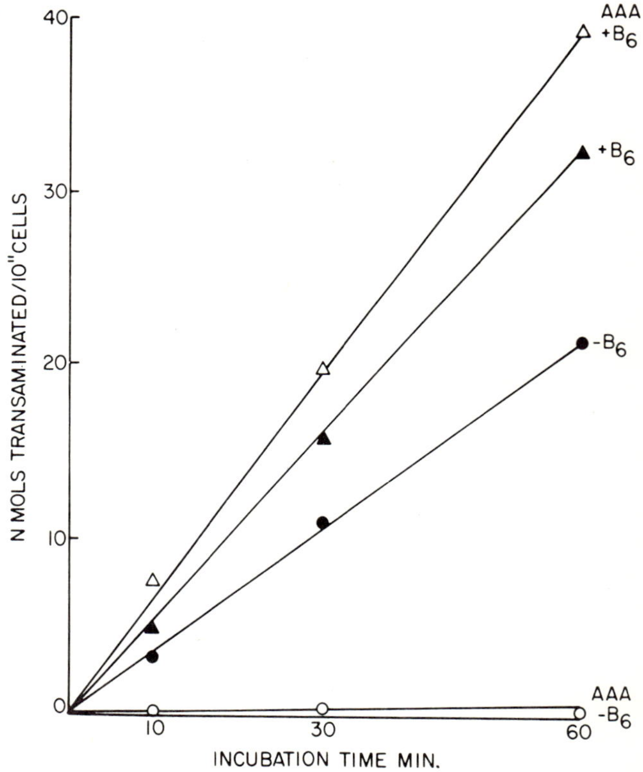

FIG. 4. Inhibition of leucine transaminase by amino oxyacetic acid. Sonified platelets were incubated with 10^{-3}M amino oxyacetic acid (AAA) for 60 min (o,Δ); then 1 nmol/ml leucine-^{14}C was added together with 200 nmols/ml α-ketoglutaric acid ± 200 nmols/ml PLP (AAA + B_6,Δ; or AAA − B_6,o). In control experiments, sonified platelets were incubated for 60 min without amino oxyacetic acid with (+ B_6,▲) or without (− B_6,●) 200 nmoles/ml PLP. Results are the mean of 2 series of observations.

TABLE 5. *Apparent K_m and V_{max} of platelet amino acid transaminases*

α-Amino acid	K_m (M)	V_{max} (nmoles/min/10^{11} cells)
aspartic acid	1.3 (10^{-4})	299
glutamic acid	4.8 (10^{-5})	45
leucine	9.1 (10^{-6})	2.2

Disrupted platelets were incubated with 0.1 to 10 nmoles/ml ^{14}C-labeled α-amino acid with 200 nmoles/ml of the appropriate α-keto acid (α-ketoglutaric acid for aspartic acid and leucine; oxaloacetic acid for glutamic acid) for 10 min at 37°C as described in Methods. 200 nmoles/ml PLP was also added. The calculations are based on observations made in six experiments.

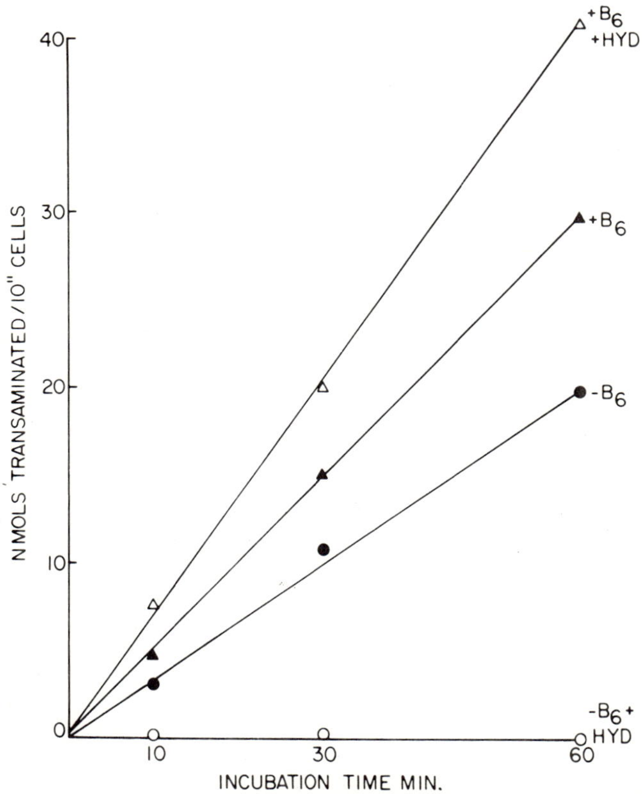

FIG. 5. Inhibition of leucine transaminase by hydroxylamine. Sonified platelets were incubated with 10^{-3}M hydroxylamine (HYD) for 60 min (o, Δ); then 1 nmol/ml leucine-^{14}C was added together with 200 nmols/ml α-ketoglutaric acid ± 200 nmols/ml PLP (Δ, + HYD + B_6; o, − B_6 + HYD). Incubation was then continued for 10–60 min as shown in the figure. In control experiments, sonified platelets were incubated for 60 min without hydroxylamine (▲, ●) and then for a further 10–60 min with PLP (▲, + B_6) or without PLP (●, − B_6). Results are the mean of 2 series of observations.

The Michaelis constants and V_{max} for all three amino acids are given in Table 5; these were obtained by using saturating concentrations of amino donor. Therefore, they are referred to as "apparent" Michaelis constants.

The role of pyridoxal phosphate in these transamination reactions was studied further by using the inhibitors amino oxyacetic acid and hydroxylamine. Figures 4 and 5 show that the transamination of leucine was completely suppressed when 10^{-3}M inhibitor was incubated with sonified platelets for 60 min prior to measurement of transamination over a further 60-min period. This inhibitory effect was overcome when the inhibitor was added in the presence of pyridoxal phosphate. Surprisingly, transamination was increased when platelets were incubated with inhibitor and PLP; further experiments will be necessary to establish the biological significance of this observation. Similar results were also obtained when aspartic or glutamic acids were used as substrates for transaminase.

IV. DISCUSSION

Previous work with blood platelet transaminase has been confined to studies with glutamic acid. The experiments that I have described show that platelets can also transaminate a number of other amino acids, some of which are of particular interest to the neurochemist because they are either precursors of neurohumoral transmitters in the brain, such as tyrosine, phenylalanine, and 5-hydroxytryptophan, or potential transmitters themselves, such as γ-amino butyric acid. Additionally, the platelet also transaminated leucine and aspartic substances whose roles in intermediary metabolism and protein synthesis are indisputable.

All these amino acids were transaminated to some degree by enzyme from disrupted cells, but with whole cells activity was confined to aspartic and glutamic acids, leucine, and tyrosine. This difference in activity could be due to differences in rate of penetration of the amino acids into the platelets. In particular, although phosphate buffer appeared to be the best medium for transamination of leucine and aspartic and glutamic acids by enzyme from sonified platelets, it may not have been the most suitable medium for experiments with whole platelets, because it was devoid of Na and Ca ions, cations which may be required for accumulation of some amino acids.

Regarding the transamination reactions with enzyme from disrupted cells, conversion of leucine, aspartic acid, and glutamic acids was unequivocal and fairly rapid, being stimulated by pyridoxal 5′-phosphate and inhibited by hydroxylamine and amino oxyacetic acid in the absence but not in the presence of pyridoxal. On the basis of these results, it seems clear that transamination of these substances is a pyridoxal-dependent process. It was interesting that the apparent Michaelis constant (K_m) for transamination of leucine indicated that this amino acid has approximately 10 times greater affinity for the enzyme than aspartic acid, although the maximal rate of transamination (V_{max}) of the latter was 100 times faster than leucine. The explanation of this difference is not apparent, but the high affinity of leucine for the transaminase may have some physiological significance. The fatty acids which can be synthesized from the product of leucine transamination, α-ketoisocaproic acid, may be involved in platelet aggregation. In this regard, ^{14}C-leucine has been used to study amino acid incorporation into protein in platelets, and only extremely low incorporation has been noted (Cooper and Firkin, 1970). This indicates the rapid transamination of leucine presumably results in incorporation of radioactivity into many non-protein fractions.

Up to this point, I have discussed only three of the seven amino acids investigated in this work. There was a firm basis for the experiments with aspartic and glutamic acids which utilized the well-known platelet enzyme, glutamic oxaloacetic transaminase; also, leucine behaved similarly, being transaminated by the above enzyme or a similar one. Unfortunately, these three amino acids are of less interest to the neurochemist than the others whose role in brain neurohumoral transmitter synthesis has already been indicated. The transamination of 5-hydroxytryptophan, tyrosine, phenylalanine, and γ-amino butyric acid was not stimulated by PLP, casting doubt on whether the reactions studied really involved enzymatic transamination. Taking the optimistic view, apart from the fact that failure to demonstrate unequivocal transamination may merely indicate that the experimental conditions were suboptimal, a possible explanation for the differences is that platelets possess specific aliphatic and aromatic transaminase enzymes. Fonnum and Larsen (1965) and Tangen, Fonnum, and Larsen (1965) have shown that there are three separate transaminases for aromatic amino acids in the brain. Consequently, there may be many separate enzymes in platelets also. Although the results do not enable us at present to consider the platelet as a model for investigation of transamination of these amino acids in the brain, it would be premature to draw the conclusion that the use of such a model will not be possible in the future when the optimal requirements for cofactors, anions, cations, keto acid amino acceptors, and substrate concentrations are established. Also, there may be induction of transaminase enzymes. In particular, the cyclic adenine nucleotide 3′,5′-cyclic adenosine monophosphate (cAMP) might cause such activation. Pastan and Perlman (1970) have described increased synthesis of various enzymes in bacteria; also cAMP or the dibutyryl derivative enhance the transport of various amino acids into bone and kidney (Phang, Downing, and Weiss, 1970) and the transport of the iodine into thyroid cells (Knopp, Stolc, and Tong, 1970). Thus, activation of platelet transaminase is an intriguing possibility.

REFERENCES

Barbeau, A. (1969): Parkinson's disease as a systemic disorder. In: *Third Symposium on Parkinson's Disease,* edited by F. J. Gillingham, pp. 66–73. Livingstone, Edinburgh.

Bartholini, G., Gey, K. F., and Pletscher, A. (1970): Enhancement of tyrosine transamination *in vivo* by catecholamines. *Experientia, 26*:980.

Black, I. B., and Axelrod, J. (1968): Elevation and depression of hepatic tyrosine transaminase activity by depletion and repletion of norepinephrine. *Proceedings of the National Academy of Science, 59*:1231.

Black, I. B., and Axelrod, J. (1969): Inhibition of tyrosine transaminase activity by norepinephrine. *Journal of Biological Chemistry, 244*:6124.

Black, I. B., and Axelrod, J. (1970): Regulation of hepatic tyrosine transaminase (TT) by norepinephrine (NE). *Federation Proceedings, 29*:736 (sbstract).

Boullin, D. J., and O'Brien, R. A. (1969): The accumulation of guanethidine by human blood platelets. *British Journal of Pharmacology, 35*:90.

Boullin, D. J., and O'Brien, R. A. (1970*a*): Accumulation of dopamine by blood platelets from normal subjects and Parkinsonian patients under treatment with L-DOPA. *British Journal of Pharmacology, 39*:779.

Boullin, D. J., and O'Brien, R. A. (1970*b*): Abnormalities of 5-hydroxytryptamine uptake and binding by blood platelets from children with Down's Syndrome. *Journal of Physiology, 212*: 287.

Cooper, I. A., and Firkin, B. G. (1970): Amino acid transport into human platelets and subsequent incorporation into protein. *Thrombosis et Diathesis Haemorrhagica, 23*:140.

Fonnum, F., and Larsen, K. (1965): Purification and properties of dihydroxyphenylalanine transaminase from guinea pig brain. *Journal of Neurochemistry, 12*:589.

Hall, Z. W., and Kravitz, E. A. (1967): The metabolism of γ-amino butyric acid (GABA) in the lobster nervous system I - GABA glutamate transaminase. *Journal of Neurochemistry, 14*:45.

Karmen, A., Wroblewski, F., and La Due, J. S. (1955): Transaminase activity in human blood. *Journal of Clinical Investigation, 34*:126.

Knopp, J., Stolc, V., and Tong, W. (1970): Evidence for the induction of iodide transport in bovine thyroid cells treated with thyroid-stimulating hormone or dibutyryl cyclic adenosine 3′, 5′-monophosphate. *Journal of Biological Chemistry, 245*:4403.

Koj, A., Zgliczynski, J. M., and Frendo, J. (1960): Distribution of transaminases in human blood cells. *Clinica Chimica Acta, 5*:339.

LaDue J. S., Wroblewski, F., and Karmen, A. (1954): Serum glutamic oxaloacetic transaminase activity in human acute transmural myocardial infarction. *Science, 120*:497.

Meltzer, H. (1968): Creatine kinase and aldolase in serum: abnormality common to acute psychoses. *Science, 159*:1368.

Pastan, I., and Perlman, R. (1970): Cyclic adenosine monophosphate in bacteria. *Science, 169*:339.

Phang, J. M., Downing, S. J., and Weiss, I. W. (1970): Cyclic AMP stimulation of amino acid uptake in bone and kidney. *Biochimica et Biophysica Acta, 211*:605.

Semba, T., and Civen, M. (1970): Subcellular distribution of aromatic amino acid transaminases in rat brain. *Journal of Neurochemistry, 17*:795.

Tangen, O., Fonnum, F., and Haavaldsen, R. (1965): Separation and purification of aromatic amino acid transaminases from rat brain. *Biochimica et Biophysica Acta, 96*:82.